The Complete Wireless Communications Professional: A Guide for Engineers and Managers

For a complete listing of the *Artech House Mobile Communications Library,* turn to the back of this book.

The Complete Wireless Communications Professional: A Guide for Engineers and Managers

William Webb

Artech House
Boston • London

Library of Congress Cataloging-in-Publication Data
Webb, William, 1967–
    The complete wireless communications professional : a guide for engineers and managers / William Webb
        p.   cm. — (Artech House mobile communications library)
    Includes bibliographical references and index.
    ISBN 0-89006-338-9 (alk. paper)
    1. Wireless communication systems.   2. Mobile communication systems.
    I. Title.   II. Series.
TK5103.2.W42   1999
621.6845—dc21   98-51802
                                                     CIP

British Library Cataloguing in Publication Data
Webb, William, 1967–
    The complete wireless communications professional : a guide for engineers
    and managers—(Artech House mobile communications library)
    1. Wireless communication systems.
    I. Title
    621.3'82

    ISBN   0-89006-338-9

Cover design by Lynda Fishbourne

© 1999 ARTECH HOUSE, INC.
685 Canton Street
Norwood, MA 02062

International Standard Book Number: 0-89006-338-9
Library of Congress Catalog Card Number: 98-51802

10 9 8 7 6 5 4 3 2 1

Contents

Preface

What is a complete wireless professional?

Since we cannot know all that is to be known of everything, we ought to know a little about everything.

Blaise Pascal (1623–1662)

Introduction

When thinking about mobile radio engineers there is a tendency to assume that the engineering function relates solely to the technical aspects of the network, such as the equipment design or the network design. That is certainly a key part of the role of a mobile radio engineer. However, increasingly engineers are required to interact with professionals from other divisions. The "complete wireless professional" should know about mobile networks; fixed networks; other types of mobile systems; regulatory and government policy; the requirements of the users; and financial, legal, and marketing issues. Otherwise, there is a tendency for the engineering, finance, and marketing departments to be completely separate entities that are unable to communicate accurately with each other. The net result is products that do not fit the marketing requirements or are not financially viable, although they may be masterpieces of advanced engineering design.

This book looks at the range of topics that complete wireless communications professionals need to understand in order to perform their task well. Clearly, above all else, they need to have an engineering knowledge of how mobile radio systems work. Such a body of knowledge is contained in many excellent text books and reference works; the intention here is to provide the salient points in each area and a guide to further reading. In other areas such as finance, the complete wireless professional only requires an understanding of the key issues, and the brief description provided in this book may be sufficient.

To some extent, this book gathers reference material from a wide range of technical, managerial, and financial sources and represents in summary form those issues that are key to the complete wireless professional. By encompassing the information in a suitable framework and providing additional chapters on areas such as the resolution of conflicts and career structure, it is hoped that the effect of the whole is greater than the sum of its parts.

Above all, complete wireless professionals need to understand the world around them and apply this knowledge to engineering issues. This book describes the world of mobile radio.

Format of this book

This book is divided into five parts.

- ▶ This first part provides introductory material in the form of a chapter discussing the relevant history of mobile communications.

- ▶ The second part looks at mobile radio systems, considering the basics of mobile radio, the design of cellular and private radio systems, and the issues concerned with interworking with the fixed network.

- ▶ The third part looks at the role of a mobile radio operator and discusses the design of mobile radio networks, the operation of these networks, the needs of large user groups, and the future of mobile radio systems.

▶ The fourth part looks at the regulatory and government decisions that impact mobile radio, including the management of radio spectrum, the standardization of mobile radio systems, and the effect of government policy on the mobile radio community.

▶ The final part focuses on becoming a better engineer by considering the resolution of conflicts such as the *time division/code division multiple access* (TDMA/CDMA) debate, on the need for understanding managerial issues, and finally on the way to become a complete wireless professional through professional vehicles and career structure.

Acknowledgments

In writing this book, I drew upon all my experience gained over years in the industry. I have learned something from almost everyone with whom I have come into contact and thank all of those with whom I have had discussions. Special thanks are due to a number of key individuals. During my time at Multiple Access Communications, Professor Ray Steele, Professor Lajos Hanzo, Dr. Ian Wassell, and Dr. John Williams, among others, taught me much about the workings of mobile radio systems. At Smith System Engineering, Richard Shenton, Dr. Glyn Carter, and Mike Shannon provided valuable knowledge, as have contacts with a number of others in the industry, including Michel Mouly (Independent Consultant), Dirk Munning (DeTeCon), Michael Roberts (Alcatel), Mike Watkins and Les Giles (Racal-BRT), Mike Goddard and Jim Norton (Radiocommunications Agency), and Phillipa Marks (NERA). At NetCom Consultants, Steve Woodhouse, Don Pearce, Mark Cornish, and others taught me about WLL and cellular operators. While at Motorola, Raghu Rau has provided invaluable guidance. Many presentations and papers from those involved in the mobile radio industry contributed to my understanding. Most importantly, my wife, Alison, has persevered as I worked many evenings and spent much time away from home in my quest to absorb knowledge in as many spheres as possible, to gain additional qualifications, and to try to become a complete wireless communications professional myself.

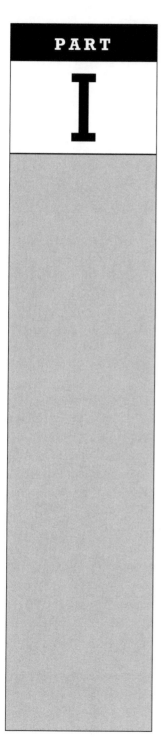

PART

I

Introductory material

CHAPTER 1

Some interesting history

When I want to understand what is happening today, or to try and decide what will happen tomorrow, I look back.

Oliver Wendell Holmes, Jr.

1.1 Introduction

The complete wireless professional does not need to be a historian. One will be able to design mobile radio networks or products equally well whether aware or not that the first demonstration of a radio transmission was made by Hertz. However, the complete wireless professional would do well to learn from some of the lessons of mobile radio over its history. Because of the dramatic change in the design and deployment of mobile radio over the last 100 years and the change in economics and uses of radio systems, the most relevant lessons are those from most recent times. Hence, this chapter provides a short summary of the historical background and tries to select some of the key lessons that can be learned. An excellent description of the history of mobile radio development, including a detailed analysis of the last 15 years of cellular deployment, can be found in [1],[1] while a very readable biography of the founder of Motorola, Paul Galvin [2], describes how Motorola shaped much of the development of mobile radio from 1930 onward. This section provides a short summary of the topics that the references describe in much greater detail.

The key issues that the complete wireless professional should learn from this chapter are:

▶ History, particularly recent history, has a number of important lessons that do not seem to have been learned by many in the mobile radio business;

▶ Standardization is far from an assured route to success, with the *global system for mobile communications* (GSM) being the only standardized product among a wide range of standards to have been highly successful;

▶ Standardization of GSM took 13 years in total;

▶ New radio systems such as cdmaOne can emerge from unexpected sources and, with the right political and economic backing, can become important globally.

1.2   Early history

Some of the key historical developments leading to today's mobile radio systems are shown in Figure 1.1.

The first radio transmissions are attributed to Heinrich Hertz in 1885. The first radio system was very simple, using a switch and an induction coil to generate a spark across two electrodes. The receiver was a loop made from copper wire around 35 cm in diameter, with a small gap in the loop. When the "transmitter" generated a spark, a small spark was seen to jump the gap in the receiving coil. Spark transmission was used as the basis for most radio equipment until around 1915 and is the reason why radio operators are sometimes known as "sparks." Hertz's work aroused

---

1. This is a superbly written and highly authoritative book and comes highly recommended.

Technical milestones		Applications
	1880	
Theoretical prediction of radio waves Generation of radio waves (Hertz) Tuned circuit (Lodge)	1890	
Aerial/earth system (Marconi)	1900	Cross-channel tests (UK) Royal Navy (UK)
Speech transmission (Fessenden) Thermionic valve (Fleming)	1910	Transmission to automobile (US) Merchant shipping (UK) Transatlantic telegraph service
Valve transmitter (Meissen)	1920	Radio direction finding (UK) Aircraft use for artillery spotting (UK) Transportables (UK)
First international spectrum conference	1930	Police use (Detroit, US) Fishing boats (Norway, UK) Aviation navigation and control
Frequency modulation (Armstrong)	1940	Telephones on ocean liners (UK)
	1950	Private mobile radio systems (US) Operator controller mobile phones (US)
Cellular concept (Bell Labs) Junction transistor (Schockley)		
	1960	
Digital integrated circuits	1970	
		Automatic mobile telephones Cellular test (US)
Solid state switches Microprocessor	1980	Cellular services (Japan)
	1990	
		Digital cellular networks (Europe) Iridium satellite services launched

Figure 1.1 Key historical milestones.

much scientific interest, but it took the emergence of a scientist with business acumen to move these discoveries into the commercial domain.

Perhaps the father of today's mobile radio systems was Guglielmo Marconi, born in Italy in 1874. Marconi was first a scientist who made the important breakthrough of redesigning the transmitter from a gap between two electrodes to connecting one electrode to the earth and the other to a metal pole placed on top of a mast (the aerial of today). By this method, he was able to demonstrate transmissions over a range of a few kilometers rather than just across the laboratory. Marconi tried to approach the Italian Post Office for sponsorship of his work but was unsuccessful. He then moved to the United Kingdom, where the British Post Office was prepared to provide sponsorship. This proved to be a shrewd move because the first application of mobile radio was in shipping and, at the time, the United Kingdom had the world's largest shipping fleet. Marconi's first sales of radio systems were in 1900 when the Royal Navy ordered 32 sets at the equivalent of today's cost of $500,000 per set with annual royalty payments of $270,000 per set. It was revenue from deals such as this that enabled Marconi to found a company and develop new radio products. For some time, the Marconi company was the world's largest producer of radio equipment. Marconi was awarded the Nobel Prize for Physics in 1907 and died in 1937. The company he founded is now part of the *General Electric Corporation* (GEC).

Other important developments included the invention of the thermionic diode in 1904, which lead to practical high-vacuum triodes by 1912, facilitating the use of narrower band transmissions and making the transmission of speech a possibility. The superheterodyne receiver was developed by Armstrong and Fessenden in 1912, and by 1933 Armstrong had developed the concept of frequency modulation.

The First World War proved an important vehicle for demonstrating the value of mobile radio in military maneuvers, especially for use by spotter planes providing reports for artillery. The need for smaller and lighter transmitters for planes hastened the reduction in size of radios to the extent that they could be carried in a backpack.

After the First World War, the main impetus for developments came from broadcasting. The rapid increase in the number of radio stations, especially in the United States, resulted in a commercialization of receivers. It also resulted in efforts to coordinate the use of radio spectrum, and the first international spectrum management conference took place in Washington in 1927. This standardized the use of frequencies up to 1.5 MHz—the highest frequency thought to be of practical use for radio transmission. Some early experiments were also undertaken in what has become known as private mobile radio. In 1921, the Detroit police experimented with voice transmission to cars, but only in a one-way format, and the Metropolitan police in London conducted a similar experiment in 1923.

By the Second World War, domestic radio receivers were relatively complex, with a high sensitivity and selectivity and the ability to receive radio stations from around the world. During this period, two-way radios slowly developed to the extent that some police forces were equipping police cars with radio transmitters, but the high power consumption and weight prohibited their universal acceptance.

The Second World War resulted in the mass production of mobile radio equipment in order to equip the increasing number of military aircraft and ships. Infantry backpack radios became more popular and around 50,000 were manufactured in the United Kingdom during the war. After the war, the manufacturers were looking for a market for their large production capability and started to target the *private mobile radio* (PMR)[2] market. Mobile phones were fitted in taxis from 1950 onward, and the basic dispatch form of communications still used today, and described in more detail in Chapter 4, was developed.

The next step forward was the development of the transistor, dramatically reducing the size and power consumption of radio systems and enabling mass production of circuit boards to reduce prices. By 1965, the first pocket-sized mobile phones were produced, allowing market growth so that, for example, every policeman could be equipped with a mobile radio rather than every car. The penetration of PMR at this time started to owe more to the licensing and regulatory policies of the government rather than the equipment or market acceptance, and even today this has resulted in a situation where the United States has more than four times the percentage penetration of PMR than the United Kingdom. Regulatory policies and the implications for the mobile radio engineer are discussed in more detail in Chapter 12.

---

2. Private mobile radio is known as *specialized mobile radio* (SMR) in the United States and as *private business radio* (PBR) in the United Kingdom, although PMR is by far the most widely used abbreviation and will be adopted throughout this book.

1.3   Some key milestones in
mobile radio history

In describing the first cellular systems it is important to remember that there is always a thin dividing line between PMR and what is today known as mobile radio, typically cellular radio systems. Both are basically transmitter/receiver units; the differences typically lie in the services with which they are equipped. This is a topic to which we will return in more detail in Section 14.2. The PMR systems described previously were typically technically able to provide some form of mobile radio service but were normally prohibited from interconnection to the telephone system (and still are today) both to prevent longer calls (which are typical of connecting to a landline subscriber), which would therefore congest the radio spectrum, and to preserve the licensed (at the time monopoly) provision of cellular radio services. The first mobile phone service was introduced by AT&T in 25 U.S. cities in 1946 and called *Mobile Telephone Service* (MTS). It was not a cellular system because only single cells were used and operator intervention was required to set up calls. This was followed by the *Improved Mobile Telephone Service* (IMTS), which also only used single cells but allowed automatic call set-up using tone signaling. However, a shortage of spectrum and a lack of government interest in correcting this situation prevented this system from providing any significant capacity and it made little impact.

In Sweden, the first European mobile radio system was introduced in 1955 by Televerket; with modifications, this system existed until around 1981, when its subscriber base had grown to 20,000 users. In the United Kingdom, the first commercial system, called System 1, was introduced in 1965 in London. It was expensive, had limited capacity and many drawbacks, but was still heavily oversubscribed. The next variant, System 2, was never deployed, but System 3 reduced the voice channel bandwidth from 100 kHz to 25 kHz, increasing the capacity. This still fell a long way short of the demand. It was not until the early 1980s that cellular mobile phone systems were deployed, finally providing the dramatic increase in capacity required to make mobile radio a mass-market product. The cellular concept is described in more detail in Chapter 2.

Much of the work on cellular systems was pioneered in the United States. The cellular concept was first developed by Bell Labs in 1948, and its parent company, AT&T, lobbied the government for radio spectrum for some time. In 1977, this eventually resulted in an assignment in the 800-MHz band, which is still one of the key frequency bands for mobile radio. Trials based on this license took place until 1981 and provided very encouraging results. The trials convinced the U.S. regulator, the *Federal Communications Commission* (FCC), that cellular was a viable concept. There then followed a long period during which the FCC tried to determine how to best assign licenses for cellular, the start of a protracted and still ongoing process of selecting the optimum way to assign licenses that is described in more detail in Section 12.3. In this case, awarding the licenses based on the quality of the application (the "beauty contest" process) failed to work due to the huge number of applicants, making evaluation highly difficult. The FCC overcame this using the lottery approach of selecting licenses at random; it proved to be a highly unsatisfactory approach that resulted in the substantial trading of licenses after the award. In fact, the United States took over seven years to award all its licenses; all the delays resulted in a high cost to the U.S. economy in foregone revenue and growth. U.S. cellular deployments were based on the *Advanced Mobile Phone Service* (AMPS)—a standard still in widespread use today. However, the AMPS standard only defined the air interface; most operators used different approaches to switching and billing, with the result that roaming between different regions in the United States, of which there were more than 90, was not possible. For a more detailed description of cellular in general and AMPS in particular, see [3].

Developments were also taking place in other parts of the world. In the Nordic countries, the Nordic Mobile Telephone at 450 MHz (NMT450) was being developed with the advantage that it would allow roaming to other Scandinavian countries. NMT900 was subsequently introduced because the capacity of the 450-MHz frequencies proved insufficient. Other European countries adopted a range of systems, some developed within the country and only used for that country. Others adopted a modified version of AMPS known as a *Total Access Communications System* (TACS) that operated in the 900-MHz band. In Europe, this stage of mobile radio development, lasting from around 1985 to 1991, was generally marked by monopoly provision. Most countries only provided a license to the existing state *Post and Telecommunication Organization* (PTO). Only the United Kingdom took the unusual steps of introducing competition by issuing two licenses and preventing the PTO from owning either of these licenses (although they were allowed to take a minority shareholding). Competition is now recognized as important in the provision of mobile radio services, and the *European Commission* (EC) mandates that members must have a competitive mobile radio environment.

1.4   Recent history

The recent history of mobile radio since 1991 has been dominated by the introduction of digital mobile radio and the attempts to standardize third generation systems. Key within this history are the roles of the GSM, CDMA technologies, and the third generation concept and the success (or otherwise) of standardization. It is this history that is probably of greatest interest to the complete wireless professional because the lessons of this period are likely to be highly relevant over the coming years.

GSM   Mobile radio since 1991 has been dominated by the GSM system. However, in 1991, it was far from clear that this would be the case. Standardization of GSM started in 1982 within the *Conference Europeenne des Administrations des Postes et Telecommunications* (CEPT)—the European spectrum management body. Although CEPT had standardized many products in the past, they were far from successful. Typically, CEPT standardization was led by engineers with little regard for commercial reality and with a desire to see their own ideas incorporated into the standard. The standards that were developed, such as X25, were often ambiguous and resulted in various national implementations, preventing interworking in the form envisaged. The track record of other standards bodies was also not good. In most cases, the development of standards took so long that national solutions had already been developed and the acceptance of the standard was low. Hence, there was little reason to suppose that GSM standardization would be successful (and indeed, little reason to suppose that other standards, such as the *digital enhanced cordless telephone* (DECT) or the *terrestrial trunked radio* (TETRA), would be successful simply because GSM was). Standardization will be discussed in more detail in Chapter 13, but an important lesson of history is that standardization of complete mobile radio systems is more likely to fail than to succeed.

The reasons for the success of GSM are varied. One key point was that standardization began early—before almost any manufacturer had started to develop their own digital mobile radio standard. This prevented different manufacturers from going different ways and ensured that there was sufficient time for the relatively slow standardization process to produce results before the product was required. The transfer of the standardization from CEPT to the *European Telecommunication Standards Institute* (ETSI) helped to produce new "rules" about the manner in which the standardization would proceed, making the standardization more practical. Another factor was that the European Commission was in the process of mandating GSM for use by cellular operators and that many European countries were coming to a dead-end with the analog system, which was expensive due to its proprietary nature and unable to provide the additional capacity required. The inclusion of manufacturers into the standards bodies was also a very important development compared to previous CEPT standardization that included only the PTOs. The fact that there was no major competing standard, particularly from the United States, was also helpful. Finally, the vision of the participants, who foresaw the increase in semiconductor complexity, and the reduction in costs helped provide a standard that was state of the art at its time of completion. A more detailed assessment of the development of GSM is provided by [1].

The progress of GSM has been far from smooth, with many delays en route. The standardization proved much more complex than originally anticipated and took a total of 13 years from the inception in 1982 to the final delivery of the Phase Two specifications with all intended features in 1995. Other complex standards such as DECT and TETRA have taken equally long periods of time. Considering that the next generation of mobile radio standards will be even more complex, the long time taken to develop the GSM standards should not be forgotten when remembering the timescales suggested by those involved in third generation systems.

GSM is now installed in well over 100 countries. New operators making a decision about the technology they should adopt often select GSM because of the competitive supply of equipment, the large base of expertise in deploying the network, and the fact that users can roam to other countries. But clearly this was not the case for the first operators, who experienced something of a chicken-and-egg problem; that is, once successful, the standard becomes even more successful, but how does it become successful in the first instance? In the case of GSM, operators in Europe were mandated to use the technology, so hence a large volume of sales was guaranteed, spurring the manufacturers to produce competitive offerings. Many have complained about this mandating. For non-European manufacturers it provided a closed market, and for operators in Europe it removed choice. The issues surrounding technology mandates are discussed in Appendix B.

It could be argued that this combination of a standard started early, a guaranteed market base, a lack of competition from other standards, and full European cooperation, not to mention the careful and skilled work of those performing the standardization, is unusual and that standards are more likely to fail than succeed. Indeed, there is much evidence that this is the case. Unfortunately, many engineers only look back into history as far as GSM and conclude that all European standards will be successful. This is probably a rather selective use of history, as the following examples demonstrate.

DSRR   After the success of GSM, the EC started standardizing digital versions of almost every possible radio system including PMR, short range, cordless, and paging. Short range systems currently have a small but steady market around the world. These systems do not use a base station but communicate directly between mobiles in a "walkie-talkie" mode (see, e.g., [4]). They are widely used in places such as building sites and department stores where a number of people work in a relatively small area. The EC decided that Europe needed a digital standard for these applications and started the *digital short-range radio* (DSRR) standardization project. The scope of DSRR rapidly grew from a simple "back-to-back" radio system to one where terminals could relay messages to other terminals and had security features and complex digital encoding. With GSM, the approach of making the phone highly complex had worked because the economies of scale allowed this complexity to be added at little cost. The DSRR standards body failed to realize that DSRR would have much lesser economies of scale; in any case, these advanced facilities were not required by the user, who valued low cost above all. The DSRR standard was completed in 1993, but no product has ever been produced to this standard. Instead, Motorola has introduced a *short-range business radio* (SRBR) based on very simple analog transmission and with the user manually selecting one of three channels. SRBR has proved very successful in meeting the market requirements for simplicity and low price.

Standards bodies are not very good at producing simple standards. With multiple parties attending the standardization, the requirements from each tends to get added to the total specification. It is difficult to restrict the capabilities with arguments about economic viability because these are hard to quantify. Complex standards can be advantageous when there is a large market but often cause the failure of a standard.

TETRA and APCO25   The European standard for PMR, TETRA may be moving down the same route. TETRA is discussed in more detail in Section 4.3, and its role in future mobile communications is discussed in Section 14.2. Yet again, the TETRA specification has proved to be highly complex, with TETRA providing many more facilities than required by most of the users. Manufacturers are countering this to some extent by building equipment that does not have all the features in the specification, but whether this will be sufficient to generate a large enough market to cover the development costs is far from clear.

The United States has a project similar to the European TETRA project known as APCO25 that is being standardized within the *Telecommunications Industry Association* (TIA) TR8 committee (standardization committees are explained in more detail in Section 13.2). APCO has very similar goals to TETRA and, like TETRA, is targeted primarily at emergency services users. The key difference between TETRA and APCO is that TETRA uses *time division multiple access* (TDMA) while APCO25 uses *frequency division multiple access* (FDMA)—access methods are explained in Section 2.4.8.

Telepoint   Another interesting lesson is that of telepoint. The cordless and telepoint application is discussed in more detail in Section 5.2 and, again, Garrard [1] provides an excellent analysis of their history. Telepoint shows that modifying a standard from its original purpose is dangerous and that the success of mobile radio is not simply borne out of a desire for anything that can communicate without wires but for a product providing particular features. After the success of cellular in the United Kingdom, the government was keen to introduce more competition, and other industry players were keen to enter the market. The cordless standard, CT-2, developed for indoor extensions to fixed lines, seemed to offer a way to meet these requirements. By deploying a large number of cordless base stations in cities, the telepoint operators thought that they could provide a service similar to cellular with the added advantage to the users that they could use the same phone in their home, communicating with their home base station. Telepoint licenses were fiercely contested in the United Kingdom and about 10,000 base stations were deployed around the country. Users, however, were not impressed by a service that could only be used in cities, that could not accept incoming calls, and where the handsets were just as expensive as cellular.

It seemed obvious to many at the time that telepoint coverage is too expensive to provide and that coverage is a critical issue to users. The telepoint operators appear to have been blinded by a desire to operate a cellular-type network and the equipment manufacturers by a desire to sell more products. Even more bizarre was the entry into the U.K. market of another operator (Hutchison Rabbit) after the first four operators had failed. Estimates are that Rabbit never managed to have as many subscribers as it did base stations. History has shown that mobile radio can be highly successful, but only if it provides the service that the subscriber wants.

CDMA and TDMA   One of the key debates of recent years has been whether *code division multiple access* (CDMA) or TDMA is the more appropriate access scheme for mobile radio. This debate is explored in more detail in Section 14.4. In fact, and little realized by many engineers, the debate has been less about the ideal access scheme and more about whether operators should select GSM or a standard developed by the U.S. company Qualcomm, now called cdmaOne (previously referred to as IS-95). The debate, although overtly technical, has really been an issue of trying to market cdmaOne as better than GSM. When cdmaOne was announced in the early 1990s, it seemed unlikely that it would succeed. The company that designed it was relatively small and little known in the world of mobile radio. Standards were already established in Europe (GSM), and the U.S. standard (digital AMPS, or D-AMPS) was supported by the key manufacturers. However, today, cdmaOne is probably the world's second standard, after GSM. Understanding how this occurred is an interesting historical lesson.

Probably, Qualcomm could only have succeeded with cdmaOne in the United States. This is one of the few countries that:

▶ Allows any manufacturer to develop a product that they can then offer as a standard (unlike Europe where there can only be one jointly developed standard);

▶ Has a large enough home market to produce good economies of scale in the case that the standard is not accepted elsewhere;

▶ Is sufficiently advanced in the development of cellular technologies that much of the rest of the world looks toward them for leadership.

Clearly, Qualcomm would not have (and has not) succeeded in Europe where the European standard was mandated. Key in the success of Qualcomm was their linkage with a number of other manufacturers and the desire of the United States to have a homegrown product rather than importing the European product and hence losing leadership in cellular. Other countries such as South Korea also rebelled against European dominance and insisted that their operators deploy cdmaOne, with equipment purchased from South Korean producers in order to encourage local industry. Qualcomm's astute use of partnerships and exploitation of the backlash against European dominance enabled cdmaOne to become a key global standard. Politics and national sensitivities are likely to play a key role in the future development of mobile radio standards.

Third generation Third generation systems are intended to be the replacement technology for existing second generation digital systems such as GSM. The concept of third generation is described in more detail in Chapter 11. Here, it is interesting to examine its progress to date. When third generation was first announced, the key attribute was the ability to provide service to all users, including cellular, cordless, PMR, and satellite, all within the same system. One of the uses was data rates up to 2 Mbps in some environments, although at the start of the standardization this was not a key requirement. During the time that the standards bodies were making little progress trying to agree on the basic structure of third generation, GSM was quietly evolving to provide service to nearly all users including PMR, satellite, cordless, and others. Suddenly, the third generation standards committees realized that their requirements had mostly already been met—with the exception of the 2-Mbps data. This now became the key requirement for third generation without any real indication that it was required by the users or that it was practical to provide given the limitations of radio spectrum.

Third generation standardization is ongoing and it will be interesting to watch the development of the standard. However, those doing the standardization seem to have trouble recognizing that the direction of the standardization might need to change. Nor do they appear to have learned the lesson from GSM that such standardization might take 13 years or more. Much of the standardization is performed by engineers who would do well to look at recent historical experience.

References

[1]   Garrard, G., *Cellular Communications: World-Wide Market Developments,* Norwood, MA: Artech House, 1997.

[2]   Petrakis, H. M., *The Founder's Touch—The Life of Paul Galvin of Motorola,* Chicago: Motorola University Press, 1991.

[3]   *Bell System Technical J.,* Vol. 58, No 1, Jan. 1979.

[4]   Walker, J., ed., *Advances in Mobile Information Systems,* Norwood, MA: Artech House, 1998.

Mobile radio systems

In this section the key fundamentals of mobile radio systems are introduced. The intention is not to write a comprehensive textbook covering all areas of mobile radio technology—many excellent books exist already—but to provide an overview. Armed with the knowledge in this part, the complete wireless professional should be able to understand the key issues and know where to look to find more detailed information if it is required.

This section assumes a basic knowledge of electrical engineering principles and some simple mathematical capability. For those who require a more basic introduction to the principles of mobile radio engineering, *Understanding Cellular Radio* by W. Webb (Norwood, MA: Artech House, 1998) is a suitable introductory text.

CHAPTER 2

Contents

The basics of mobile radio

Man's business here is to know for the sake of living, not to live for the sake of knowing.

Frederic Harrison

2.1 Introduction

The basic principles of mobile radio are best understood by first studying the propagation mechanisms by which the signal passes from the transmitter to the receiver. From propagation, this section examines the shortage of radio spectrum and the complex system designs required to provide sufficient system capacity. Then the design of a typical system is examined, providing a good understanding of the basics of mobile radio communications. The complete wireless professional needs to have an understanding of:

- The likely received signal strength and the effect of slow and fast fading and *intersymbol interference* (ISI) on the received signal (these terms are explained later in this chapter);

- The lack of radio spectrum and its implications for mobile radio system design, including the means whereby spectrum can be reused;

- The overall block-diagram design of a mobile radio system and the need for each of the blocks;

- The capacity (i.e., the number of subscribers) that can actually be achieved from mobile radio systems.

This section provides a basic guide to these issues.

2.2 Basic principles of propagation

An understanding of radio propagation is essential to the complete wireless professional because the loss in signal caused by propagation limits the received signal strength, impacting on the quality of the received signal. Many of the building blocks of mobile radio systems, as introduced in Section 2.4, are used solely to overcome the problems introduced by propagation. These building blocks include error coding, equalization, and to some extent the choice of multiple access scheme. There are many detailed treatises on propagation, and the topic is covered in a wide range of books such as [1–3], in most cases in a highly mathematical fashion. Here the key issues are introduced.

If the received signal strength at a mobile radio is plotted against time, then the trace would show a great deal of complexity that would typically take substantial effort to understand. To simplify the analysis of radio propagation, engineers generally consider the received signal strength to be a composite of three discrete effects known as path loss, slow fading, and fast fading. Although such characterization does not exactly reflect reality, it has proved to be sufficiently useful and accurate to model mobile radio systems and is in widespread use to date. Each of these separate elements is now examined.

Path loss Path loss is the simplest of all the propagation mechanisms to understand and reflects the fact that the signal drops as the distance from the transmitter increases. Theory shows that if the transmitter were in "free space" (i.e., some distance away from any object), then the signal would radiate in an expanding sphere from the point source of the transmitter. Since the surface area of the sphere is proportional to the radius squared, the received signal power at a distance d from the transmitter is proportional to $1/d^2$. Free-space loss cannot occur on the Earth since one half of the expanding sphere is under the ground which has a certain reflection and transmission coefficient depending on the material making up the surface of the Earth at that particular point. Of more relevance is the fact that there will be obstructions on the ground in the form of buildings, hills, and vegetation, for example. These absorb and reflect the signal, resulting in a received signal strength that is much lower than that predicted using free-space loss. Because of the complexity of modeling every building, general guidelines are adopted as to the loss likely to be experienced. Measurements have shown that in an urban environment, if the path loss is modeled as being proportional to $1/d^{3.5}$, or in some cases $1/d^4$, then the results achieved best reflect real life. Empirical models, such as that from Hata, introduced in Chapter 7, take the analysis one stage further by modifying the exponent according to the height of the mobile antenna, with the exponent falling by around 0.6 for each order of magnitude increase in the height of the mobile antenna. This reflects the fact that as the mobile antenna rises, buildings and other obstructions have increasingly less effect and, hence, the path loss can come closer to free space.

There are a few isolated cases where path loss exponents lower than the exponent of 2 predicted by free space are experienced. These typically occur in constrained spaces, often in corridors in a building. Here, the signal does not expand on the surface of a sphere because the walls of the corridor cause the signal traveling toward them to be reflected back into the corridor. Because the signal is now moving forward on a surface that is not expanding (assuming the corridor stays the same width and height), theory would predict that no loss in signal strength will occur. In practice, some signal leaks through the corridor walls and exponents of around 1.6 to 1.8 can be realized.

Slow fading The word *fading* is used to describe a drop in the received signal strength, over and above that which would be expected based upon path loss. This loss occurs temporarily. There are two phenomena in mobile radio that cause fading to occur, one that causes fades lasting of the order of a few seconds and one causing fades lasting of the order of a few microseconds. The former is termed slow fading, while the latter is termed fast fading. To see a slow-fading waveform it would be necessary to take a set of measurements made by a mobile; to remove the effects of path loss by correcting for the distance from the transmitter at any point, using a formula such as that determined by Hata; and then to filter the remaining signal such that any high-frequency changes, lasting less than a second, were removed. The resulting waveform would typically show a signal falling by around 8 dB or so over a period of a few seconds and then rising back up to the mean level.

This fading phenomenon is caused by the receiver temporarily passing behind obstacles that partially block the signal from the transmitter. A clear example of this is realized when driving down a street that has detached houses between the base station and the mobile. When behind the houses the signal strength will be reduced, whereas when between them the signal strength will rise back to the expected level. The depth of the fade will depend on both the amount of loss of the signal in passing through the building and the strength of signals received by other mechanisms such as reflection. The duration of the fade will depend on the time it takes the mobile to traverse the building.

Measurements have shown that if a plot of the slow-fading waveform is periodically sampled and the probability of any particular level of signal strength plotted, then the results will follow a log-normal distribution (i.e., a normal distribution plotted on a logarithmic scale) with a standard deviation of around 8 dB.

Fast fading   Fast fading can easily be seen on the plot of received signal strength if only a small portion of the plot, of duration say 1 sec, is examined. Alternatively, it can be shown by filtering out all low-frequency changes in the received signal. The cause of fast fading is multipath propagation. In a complex urban environment, the mobile will receive many copies of the transmitted signal each traveling via a different path. Some will come direct to the mobile, others will reflect off buildings, cars, or other objects. The simplest arrangement of a mobile receiving a direct and a reflected signal from the transmitter is shown in Figure 2.1. In real life, the receipt of a direct signal is relatively rare, and the mobile is likely to receive numerous reflections from all around. Each of these signals has followed a different path from the transmitter to the receiver and so is likely to have a different signal strength. The length of each path will also

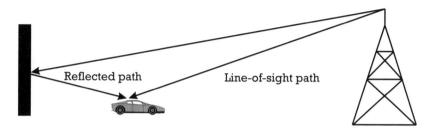

Figure 2.1   Multipath propagation.

be different, with the result that each wave will take a slightly different time to arrive at the mobile. This will result in the phase of the carrier wave being different for each of the received waveforms.

Imagine the simple case where there are two received waves, both of the same signal strength, and that one results from a reflection of a moving object, so that over time the distance this second wave travels increases. The result is shown in Figure 2.2, where the top trace shows the first wave, the second trace the reflected wave, and the third trace the composite signal as seen by the receiver. It is clear that at the point at which the waves are in exact antiphase there is complete cancellation of the received signal, resulting in a fade that extends to zero received signal strength. At a transmission frequency of 900 MHz, the transition from constructive interference to destructive interference as shown in Figure 2.2 would take half a wavelength, approximately 15 cm. Hence, each time the mobile moves a full wavelength, around 30 cm, it is likely to pass through a fade. A mobile will travel this distance in a very short period of time—hence, the term *fast fading*.

This can alternatively be represented by a process of vector addition as shown in Figure 2.3, where the magnitude of the vector represents the strength of the signal, and the angle of the vector from the origin (in an anticlockwise direction) represents the phase difference between the "reference" signal (the strongest path) and the particular received signal. There can be a number of vectors each corresponding to a different received path. In the case that the received signals are exactly in phase (as is the case at the start of Figure 2.2), the vector diagram consists of two vectors superimposed on top of each other. In the case that the vectors are in exact antiphase (as is the case at the end of Figure 2.2), the two vectors are of equal magnitude but opposite polarity. Clearly, addition of the two vectors in the first case would result in a single vector along the origin of

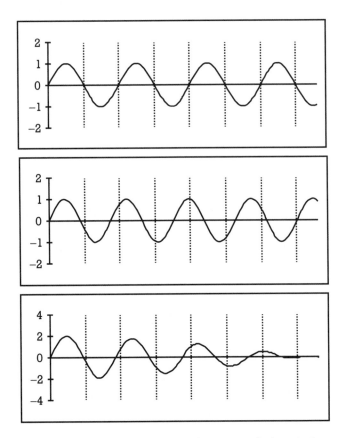

Figure 2.2   The top trace represents the wanted signal, the second the interfering signal, and the third the received waveform.

twice the magnitude, whereas in the second case the result of the addition would be no remaining signal. Figure 2.3 shows the vector diagrams and resulting addition for the cases of phase differences of 90 degrees and 180 degrees.

It real life, the prospect of exact cancellation is very remote since no two waves will typically have the same signal strength. However, the probability of a partial cancellation is very high, with the result that the signal will fluctuate by as much as 40 dB from the mean level while the mobile passes through fades. Figure 2.4 shows what a typical fast fading signal would look like. Fast fading is also known as Rayleigh fading, after the physicist who developed the statistics that can be used to describe it.

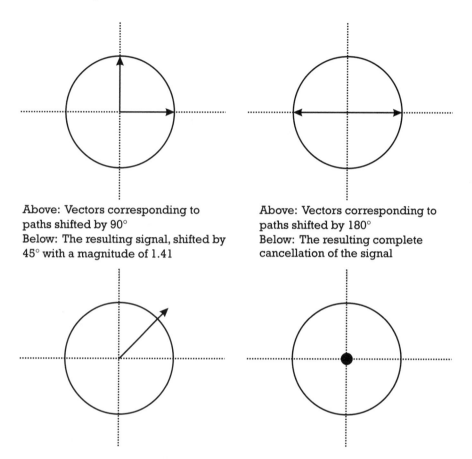

Above: Vectors corresponding to paths shifted by 90°
Below: The resulting signal, shifted by 45° with a magnitude of 1.41

Above: Vectors corresponding to paths shifted by 180°
Below: The resulting complete cancellation of the signal

Figure 2.3 Vectorial representation of multipath signal reception.

Mathematically, Rayleigh fading is modeled by the combination of in-phase and quadrature signals, both having a normal distribution with variance σ^2. Then the amplitude of the received signal is given by

$$a(k) = \sqrt{a_i^2(k) + a_q^2(k)} \qquad (2.1)$$

where a_i and a_q are the in-phase and quadrature waveforms, respectively. The mathematics of deriving the distribution of the amplitude are complex and detailed in [4]. In the case that there is no dominant path to the mobile, the probability of the signal having a particular amplitude a is given by

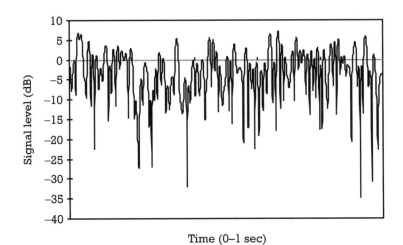

Figure 2.4 Rayleigh fading waveform for a mobile moving at walking speed over 1 sec.

$$p_{\text{Rayleigh}}(a) = \frac{a}{\sigma^2} e^{-a^2/2\sigma^2} \tag{2.2}$$

A graph of this equation for the case $\sigma = 1$ is shown in Figure 2.5, where it can be seen that the highest probability is for the signal to have no fading; that there is a significant probability of fades as deep as 22 dB; and that the tail of the graph asymptotically approaches the x-axis, resulting in a small probability of an infinitely deep fade.

If there is not a *line-of-sight* (LOS) path between the transmitter and the receiver, and typically there is not, then there are only two mechanisms by which the radio signal can propagate from the transmitter to the receiver, namely, reflection and diffraction. At the frequencies used for a

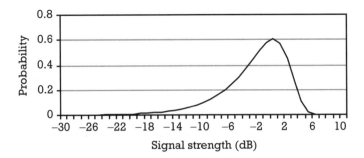

Figure 2.5 Probability of a Rayleigh fading signal strength.

mobile radio system, diffraction results in very high levels of path loss; hence, reflection is by far the most important phenomenon.

Reflection simply results from the signal reflecting from the surfaces that it encounters rather than being absorbed by them. The amount of reflection depends on the reflective, absorptive, and transmissive characteristics of the surface the wave encounters. Sheet metal provides near-perfect reflection, while glass and paper provide near-perfect transmission. Materials such as brick result in some reflection, some absorption, and some transmission. Typically, the surfaces encountered in real life are rough; hence, any reflections are diffuse, spreading the signal over a larger area but resulting in a lower signal strength.

Diffraction allows signals to bend over the edge of obstacles. Modeling diffraction is highly complex and typically is only tractable for the idealized "knife-edge" case where the obstacle encountered can be considered to have minimal width [1, 2]. The key issue with diffraction is the angle through which the signal can diffract. When a large angle of diffraction can be achieved, it is possible for the signal to "re-form" after passing around an obstruction. With lower angles, a "shadow" is formed behind the obstruction. For the knife-edge, Figure 2.6 shows the variation of signal strength of a diffracted signal with parameter v, while Figure 2.7 shows how the signal strength for a given diffracted angle varies with frequency.

The parameter v is given by

$$v = h\sqrt{\frac{2(d_1 + d_2)}{\lambda \cdot d_1 d_2}}$$

(2.3)

where d_1 is the distance from the transmitter to the obstruction, d_2 the distance from the obstruction to the receiver, h the height of the obstruction, and λ the wavelength of the transmitted frequency. For a possible GSM network deployment in a city with $d_1 = 1,000$m and $d_2 = 200$m, at 1,800 MHz the parameter $v = 0.28h$. Hence, using Figure 2.6, if $h = 0$ (i.e., the LOS path grazes the top of a building) the diffraction loss will be around 5 dB, while if the height is 10m (corresponding to a total angle of diffraction of 3.43 degrees) the loss will be around 22 dB. A little further analysis soon shows that diffraction angles of greater than 1 degree are likely to result in insufficient signal strength at the frequencies of interest. As Figure 2.7 shows, the loss is frequency dependent, but as the loss is

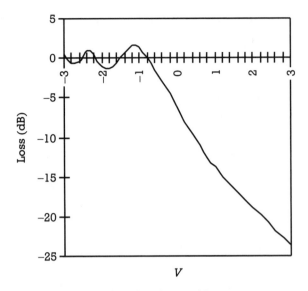

Figure 2.6   Variation of diffraction loss with parameter *v*.

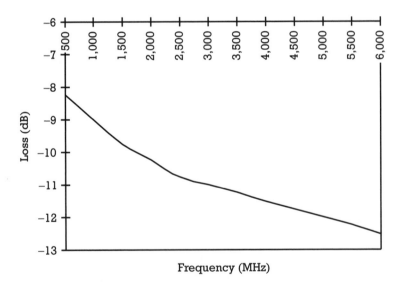

Figure 2.7   Variation of diffraction loss for a particular obstruction with frequency.

already severe, the frequency variation is unlikely to be the overriding issue.

Wideband channels   There is a further difficulty that may be caused by fast fading. If the reflection comes from some faraway object, then the delay of the reflected signal will be large. If this delay is greater than the time taken to transmit a bit of information, then when the reflected signal finally arrives it carries different information to the direct signal. The result of this is that the previous bit transmitted, or symbol, interferes with the current symbol, generating a phenomena known as ISI. ISI is problematic in mobile radio systems that have transmission bandwidths greater than around 100 kHz. In order to understand whether the problem might occur, consider the following example.

Assume that the symbol rate is 100 kHz (broadly this means that the bandwidth of the transmitted signal will be around 100 kHz). The carrier frequency is not a relevant factor in calculating ISI. (The symbol rate can be assumed to be equal to the bit rate for the purposes of this discussion.) If the reflection is delayed by 1/100 kHz, namely 10 μs, then the reflected signal will arrive during the next symbol period causing ISI. In 10 μs, radio signals travel 3 km. Hence, if the reflection occurs from an obstacle 1.5 km past the receiver, ISI will occur.

As the symbol rate increases, the distance required for ISI to occur reduces proportionally. With a bandwidth of 1 MHz, the distance falls to 300m; and at 10 MHz, down to only 30m. Some of the *wireless local loop* (WLL) systems discussed in Section 5.3 have bandwidths in excess of 1 MHz, and hence ISI might be expected to be problematic.

To further explain ISI, consider Figures 2.8 to 2.10. In each case the figures show three lines. The top line is the signal received directly by the mobile, where in this case a 010101... data stream has been sent. The second line is the signal received with some delay, although here it has been assumed that it has been received with the same signal strength. The bottom line is the composite from the first two signals, that is, the signal that the mobile radio actually receives.

In Figure 2.8, the received signal in the case that the delay is only one-fourth of a bit period is shown. The resultant signal is still clearly distinguishable as a 010101... transmission; however, the length of time that the trace spends at the level corresponding to 1, or to 0, is reduced, increasing the importance of accurate timing within the receiver. In this case, the interference would be judged to be Rayleigh rather than ISI because the delayed signal was delayed by less than a bit period. In Figure 2.9, the delay has been increased to three-fourths of a bit period. At this point, the interference is verging upon becoming ISI. Here, the

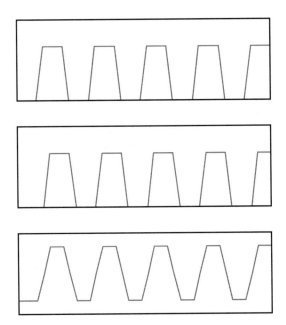

Figure 2.8   Received signals in the case that the delay is only one-fourth of a bit period.

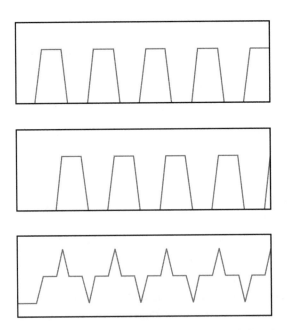

Figure 2.9   Received signals in the case that the delay is only three-fourths of a bit period.

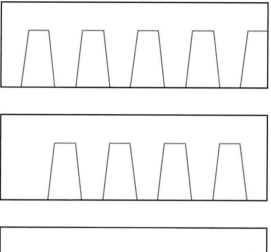

Figure 2.10   Received signals in the case that the delay is a complete bit period.

010101... signal could be recovered, but only with a perfectly synchronized receiver.

Finally, Figure 2.10 illustrates the case where the delay has become exactly a bit period. In this case, it has clearly become impossible to distinguish the transmitted information unless some additional intelligence is employed in the receiver. Now imagine the case that the transmitted data signal were 0011001100.... In this situation, it would actually be possible to distinguish the transmitted data until the delay was equal to two bit periods. Hence, ISI has the effect of rendering some transmitted data sequences unreadable, but not others. For this reason it is sometimes called *frequency selective fading* because if the transmitted data contained a range of frequencies and the frequency content of the received data was examined (e.g., using a spectrum analyzer or conducting a Fourier transform), then the received signal would appear to be missing certain frequencies (such as the frequency of the 101010... pattern in our example). ISI can be corrected through the use of equalizers, which are discussed in more detail in Section 2.4.4.

2.3 Radio spectrum utilization

Radio spectrum is the one fundamental requirement for cellular and wireless communications. All radio communications require radio spectrum, the amount required being approximately proportional to the bandwidth of the information to be transmitted. It was once said, "Spectrum is like real estate—they just don't make it any more." This is quite an apt description. Spectrum is quite like land—there is only a limited supply of it and some parts are more valuable than others. For example, certain parts of the radio spectrum have characteristics that particularly suit cellular radio, and these parts of the spectrum are in much demand from all the companies who would like to be cellular radio operators.

Radio spectrum is usually managed by the government or their agencies—for example, the FCC in the United States. It is their role to ensure that the rights to use the spectrum are given out fairly to all those who need it and to make sure that no two people are given the same bits of spectrum. Giving out radio spectrum fairly is a difficult task. It is a little like trying to fairly distribute the welfare budget; everyone seems to have a valid claim and there is not enough to go around. Recently, some governments have resorted to selling the rights to the spectrum on the basis that the person who is prepared to pay the most must be the person who needs the spectrum most badly. When enough cellular spectrum for around six cellular operators was auctioned in the United States in 1997, the government received $20B in revenue.[1] This shows just how scarce the spectrum is and how much people are prepared to pay for access to it and is a topic to which we will return to in more detail in Chapter 12.

What has tended to happen in most countries is that certain parts of the radio spectrum, such as the 900- and 1,800-MHz bands, have been set aside for cellular. This, and some other parts of the spectrum, have then been divided typically between three or four different cellular operators, with the net result that each operator has been given something like 25 MHz of radio spectrum each, normally partitioned into an uplink and a downlink band, and so the assignment is written as 2×12.5 MHz. The shortage of radio spectrum impacts upon most of the design issues of both

---

1. Actually, to be more accurate, it received pledges for over $20B. Subsequently, some of the largest bidders have defaulted on their payments with the result that the amount actually received is much less. We return to this issue in Section 12.3.

the mobile radio technology and of the network and is a critical point to be understood by the complete wireless professional.

To understand whether there is sufficient spectrum it is important to understand how the capacity of a cellular system is calculated. We first discuss the underlying theory. The capacity of a cell in terms of the number of radio channels is given by

$$m = \frac{B_t}{B_c K}$$

(2.4)

where B_t is the total bandwidth (spectrum) assigned to the operator, B_c is the bandwidth required per call, and K is the cluster size or reuse factor. Assuming that the operator has been given a fixed spectrum assignment and that they have selected a technology with a fixed bandwidth per call (as is typically the case), the remaining variable is the cluster size. The concept of a cluster is best understood with the aid of Figure 2.11. First a few words about this figure. You will note the cells are shown as hexagons. This is something of a joke in the cellular industry where nobody has yet seen a hexagonal cell in real life. Cells in real life are more or less circular (because of the way radio waves propagate, as explained

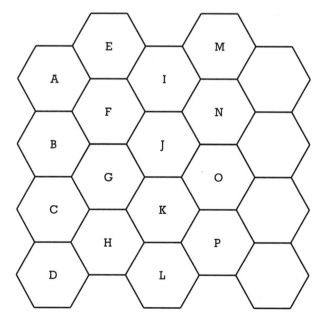

Figure 2.11 A cluster of cells.

previously), but unfortunately circular cells do not tessellate.[2] A hexagon is quite close in shape to a circle and does tessellate, and so most cells are drawn as hexagons to ease understanding.

In Figure 2.11 cells are named A to P. Now imagine that it was decided to use frequency 1 in cell A. The same frequency certainly cannot be used in cells B, F, or E because they adjoin cell A, and so there would always be interference at the edge. Typically, it cannot be used in cells C, G, J, or I either because the interference there will still be too great. However, it could be used again in cells D, H, K, O, N, or M. If we assign our frequencies on this basis, the result might start to look like Figure 2.12. Because only seven frequencies are required, it is said that there is a cluster size of seven.

Like the cluster size, the other important concept is to understand the number of radio channels required in a particular cell. Typically, the marketing department will determine how many subscribers they expect and the number of call minutes per busy hour that they expect the subscribers to make. Based on this information, the engineers need to determine how many radio channels will be needed to meet this demand. This concept is discussed in more detail in Chapter 7.

Traffic levels are turned into radio channels via the Erlang formula. Erlang was a Swedish engineer working for the Swedish Post and Telecommunications organization who worked on traffic levels in fixed networks, but his work is equally applicable to mobile networks. Erlang developed two formulae relevant to mobile radio systems. The Erlang B formula is given by

$$P_B = \frac{A^N/N!}{\sum_{n=0}^{N} A^n/n!}$$

(2.5)

where P_B is the probability of blocking, A is the offered traffic in Erlangs, and N is the number of traffic channels available. One Erlang is one continuously used traffic channel, so during any given period (say one hour) if a user talks for half the time they would be said to generation 0.5E. If there are 10 users all talking for half the time, the total traffic load would be 5E. The Erlang B formula applies to nonqueuing, or blocking, systems.

---

2. Shapes are said to tessellate when they can be placed together on a flat surface without either overlapping or leaving any gaps.

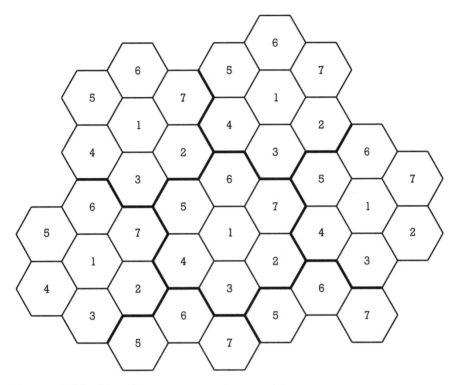

Figure 2.12 The cluster repeated many times.

That is, when a user wants to make a call, if there are no channels available, then their request to make a call is blocked. This much is true of cellular systems. However, the typical blocked user will immediately press the send button again in the hope of getting a radio channel this time, and this retry behavior is not allowed by Erlang B. Despite this, Erlang B provides results that are very close to real life for mobile radio systems and is widely used. It is normal to apply the Erlang formula to the traffic levels experienced during the busiest hour of the day, or "busy hour," in order to size the network for the worst case.

The number of traffic channels, N, required for a given A and P_B can only be solved iteratively by substituting values of N in the preceding equation until the desired P_B is reached. As a result, tables of Erlang formulae are available (a useful engineering tool is a spreadsheet macro performing Erlang B calculations that can be quite simply written—an Excel macro to perform this calculation is provided in Appendix A). A graph showing the variation of radio channels required with Erlangs of traffic and different blocking probabilities is shown in Figure 2.13. Those wishing to know more about the Erlang formula should consult [5].

There are a number of features to note about this figure. The first, unsurprising fact is that as the required blocking probability gets lower, more radio channels are needed to handle the same amount of traffic. For a typical mobile radio system, blocking probabilities of around 2% are considered acceptable, whereas for WLL systems blocking probabilities below 1% are necessary.

The second is the slight steplike nature of these curves that is caused by the fact that only an integer number (i.e., whole number) of channels is possible.

The third is that the number of channels required tends toward the number of Erlangs for high levels of traffic (e.g., using the 2% blocking common in mobile radio systems, for 1 Erlang, 5 channels are required—an efficiency of 20%; whereas for 19 Erlangs, 25 channels are required—an efficiency of 76%). This is known as "trunking gain"—the more traffic that can be trunked together, the more efficient the system.

Erlang also looked at the case where calls could be queued, as can happen in some fixed networks. In this case, he derived the Erlang C equation, given by

$$P(\text{delay}) = \frac{(pN)^N p_0}{(1-p)N!} \tag{2.6}$$

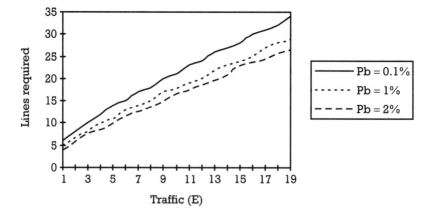

Figure 2.13 Channels versus Erlangs for a range of blocking probabilities.

where p is the utilization per trunk, given by the percentage of time the trunk will be utilized during the busy hour; N is the number of trunks; and p_0, the probability that there are no calls in the queue waiting to be served is given by

$$\left[\sum_{n=0}^{N-1} \frac{(Np)^n}{n!} + \frac{1}{1-p} \frac{(Np)^N}{N!} \right]^{-1}$$

(2.7)

Note here that instead of the probability of blocking, all calls are queued until they are served, so there is a probability of delay. In such queuing networks, the grade of service is normally specified in the form "95% of all calls must experience a delay of less than 1 sec." Erlang C does not describe mobile radio systems well because typically call requests are not queued. To date, there has not been a well-publicized formula that exactly describes the mobile radio system taking into account repeated pressing of the send button; hence, Erlang B remains the standard formula in use.

2.4 Basic system design

2.4.1 System overview

In this section, the general design philosophies of radio systems are explained. Each of the areas covered is highly complex and there will typically be a number of specialized books and journal articles covering the relevant issues. Here, an overview of the different areas is provided, along with details of where further information can be discovered. Figure 2.14 shows a block diagram of a stylized *digital* radio system. Note that the ordering of blocks is important: speech coding must be performed first, modulation and medium access last, and ciphering after error correction. Analog radio systems broadly do not have any of these steps and are sufficiently simple that they do warrant further discussion here.

Each of the stages is now described in more detail.

2.4.2 Voice encoding

Any waveform, whether it is speech or from a different source, can be converted from analog to digital by simply using an analog-to-digital

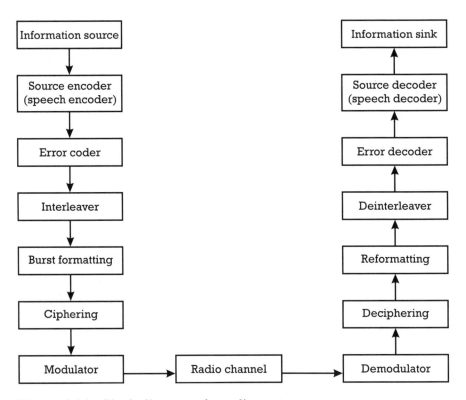

Figure 2.14   Block diagram of a radio system.

converter. All such converters sample the incoming analog signal at periodic intervals and quantify the strength of the signal at that point. The quality of signal representation in an analog-to-digital converter is measured by the number of samples per second and the number of bits used to quantify the signal at that point—the higher each of these values, the more accurately the signal will be represented. However, equally, the higher both of these parameters, the greater the information generated and the more bandwidth required to transmit the signal. Hence, a balance needs to be struck between quality of representation of signal and bandwidth required for transmission.

In all cases, there will be an error when the speech is recreated due to a difference between the digital waveform and the original analog waveform. This difference is known as the "quantization error" and is shown in Figure 2.15.

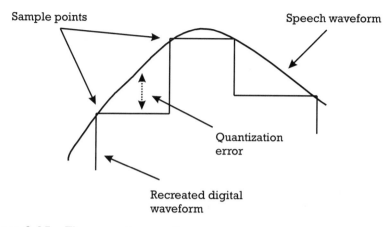

Figure 2.15   The speech sampling process.

The difficulty resides in knowing what quantization error can be tolerated. The hearing process is complex and the ear is very sensitive to some errors while very tolerant of others. Mathematical measures of errors such as the *root mean square* (RMS) value of the quantization error tend to be poor predictors of the speech quality as perceived by the user. Instead, subjective listening tests are used, where a panel of users is asked to rate the relative quality of a number of different speech coders, normally on a scale of 1 to 5. The average score is used to rate the speech coder and is known as the *mean opinion score* (MOS). Hence, the process of the design of speech coders becomes one of designing new coders and then subjecting them to panel judgment.

In the case of the simple coder described previously, the sampling rate should be set using the Nyquist theorem. This is a theorem that shows if a waveform is sampled at twice its highest frequency sine wave component, then using a knowledge of the amplitude and frequency of the various components of the signal it would be possible to rebuild the original signal. Speech contains frequencies up to around 15 kHz depending on the speaker; however, most of these higher frequency components form part of unvoiced consonants and can be "dropped" with only limited loss of intelligibility. Indeed, the key components of a speech waveform reside below 4 kHz. Hence, a sampling rate of 8 kHz is typically sufficient for an acceptable voice quality (at least equivalent to existing telephone systems).

Early speech coders used an agreed format. They took an analog-to-digital converter set at a sampling rate of 8 kbps and sampled 8 bits each time, resulting in a total data flow of 64 kbps. The quantization levels for each sample were set in a slightly nonlinear fashion where they were compressed at low signal levels and expanded for higher signal levels because experimentation has shown that the ear is more subjective to errors at low signal strength than at high signal strength. This speech-coding technique is known as "μ-law PCM." PCM speech coding is widely used in fixed telephone networks where bandwidth is not so critically constrained as for mobile networks and where high quality is considered essential. Incidentally, this implies that there is generally little point in producing a voice coder of higher quality than PCM since most calls will pass through a fixed network at some point where their voice quality will be reduced to PCM levels.

PCM is not an efficient way to encode speech. Speech waveforms when considered over a short duration of time (e.g., 20 ms) are highly repetitive and can be clearly seen to be composed of the superposition of a number of sine waves. By taking account of the predictability and perio-dicity in short samples of speech, near-identical quality can be achieved to PCM coding but at much lower levels of transmitted information. The most advanced speech coders try to model the manner in which the speech is generated. The speech coders used by digital cellular systems fol-low this route to a greater or lesser extent and are known as parametric encoders because they encode the basic parameters of the speech wave-form [1, 6].

In order to model human speech it is necessary to understand some-thing about the manner in which it is generated. Human speech can be considered as consisting of two different types of sound—voiced sounds (vowels) with a regular, periodic structure and unvoiced sounds (conso-nants) with a more noiselike characteristic that is less predictable. Exam-ples of voiced and unvoiced waveforms are shown in Figures 2.16 and 2.17. The first task of a parametric encoder is to decide whether each segment of speech is voiced or unvoiced. It then needs to determine:

▶ The equivalent filter coefficients that characterize the vocal tract of the speaker;

▶ The loudness of the speech;

▶ If voiced, the pitch information for the speech.

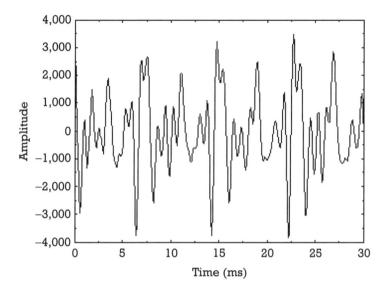

Figure 2.16 An example of voiced speech.

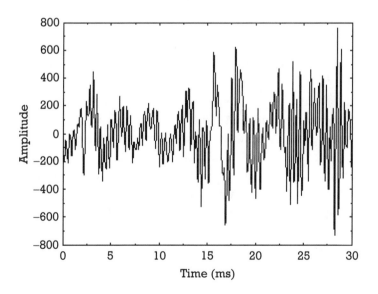

Figure 2.17 An example of unvoiced speech.

The speech decoder then recreates the speech depending on whether it was voiced or unvoiced in the following manner:

▶ In the case of voiced speech, a periodic waveform is created with the same pitch and loudness as the original and passed through a filter with the appropriate coefficients.

▶ In the case of unvoiced speech, random noise with appropriate loudness is input into a filter with the appropriate coefficients.

In the basic form described, parametric coders are very efficient at compressing speech to as little as 2.4 kbps. However, there are two problems that result in the speech quality being relatively poor:

▶ Because speech is determined to be either voiced or unvoiced, in the transition period between the different types of speech the quality will be poor.

▶ Interactions between the sound source and the vocal tract are ignored, resulting in a poor characterization of the speech in some cases.

A solution to these problems is to use hybrid coders, also known as *analysis by synthesis* (AbS) coders. Instead of simply assuming that the excitation is either a periodic waveform or white noise, they transmit more detailed information about the excitation either using a model of the pulse shape, size, and spacing or by selecting one of a number of excitation models from a code book. By selecting the best possible excitation sequence, the aim is to minimize the difference between the encoded and the decoded signals. A generic block diagram of an AbS speech coder is shown in Figure 2.18.

AbS codecs divide the input speech into frames, typically about 20-ms long. For each frame, parameters are determined for a synthesis filter, which attempts to synthesize the vocal tract, and then the excitation to this filter that most closely reproduces the speech signal is determined. This is achieved by determining the excitation signal that when passed into the determined synthesis filter minimizes the error between the input speech and the reconstructed speech. Hence the name *analysis-by-synthesis*—the encoder analyzes the input speech by synthesizing many different approximations to it. Once the analysis of the frame is completed, the encoder transmits information representing the synthesis filter parameters and the excitation to the decoder. At the decoder the given excitation is passed through the synthesis filter set with the

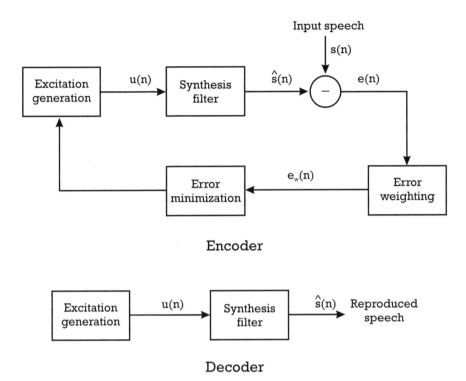

Figure 2.18 Generic model of an AbS encoder.

transmitted parameters to give the reconstructed speech. The synthesis model may also include a *long-term predictor* (LTP) that models the pitch in voiced speech, allowing the filter to shape the excitation signal more accurately.

The error-weighting block shown in Figure 2.18 is used to shape the spectrum of the difference between the input speech and the synthesized speech signal in order to reduce the perceived effect of this error. This can be achieved by noting that in frequency regions where the speech has high energy any error will be partially masked by the speech, whereas in the frequency regions where there is low energy any error will be highly noticeable to the listener. The error-weighting filter emphasizes the noise in the frequency regions where the speech content is low prior to the error minimization function, concentrating any errors into frequency regions where they will be least noticeable. Such weighting has been found to produce a significant improvement in the subjective quality of the reconstructed speech for AbS codecs.

There are many different types of AbS codecs depending on how the excitation waveform for the synthesis filter is chosen. The perfect AbS codec would pass every possible waveform through the filter to determine the excitation sequence producing the best possible match to the input speech. However the numerical complexity involved in passing every possible excitation signal through the synthesis filter is typically intractable. For most codecs, some means of reducing this complexity, without compromising the performance of the codec significantly, is used.

There are broadly three different classes of AbS codecs known as *multipulse excited* (MPE), *regular pulse excited* (RPE), and *code excited linear prediction* (CELP). The differences between MPE, RPE, and CELP codecs arise in the manner in which the excitation signal is generated.

With MPE codecs, the excitation sequence is given by a fixed number of pulses for every frame of speech. The positions of these pulses within the frame and their amplitudes are determined by the encoder and transmitted to the decoder. Determining the optimal position and amplitude for each pulse is not computationally tractable; hence, a suboptimal method of finding the pulse positions and amplitudes is used. Typically, about 4 pulses per 5 ms are used, leading to good-quality reconstructed speech at a bit rate of around 10 kbps.

The RPE codec also uses a number of pulses to characterize the excitation signal. However, with RPE codecs the pulses are regularly spaced at some fixed interval and the encoder needs only to determine the position of the first pulse and the amplitude of all the pulses. Therefore, less information needs to be transmitted about pulse positions; so for a given bit rate, the RPE codec can use more pulses than MPE codecs. For example, at a bit rate of about 10 kbps, around 10 pulses per 5 ms can be used in RPE codecs, compared to 4 pulses for MPE codecs. This allows RPE codecs to give slightly better quality reconstructed speech quality than MPE codecs. However, they also tend to be more complex.

Although MPE and RPE codecs can provide good-quality speech at rates of around 10 kbps and higher, they are not suitable for rates much below this due to the large amount of information that must be transmitted about the excitation pulses. In order to go below this bit rate it is necessary to use a CELP coder. With CELP, a large number of potential excitation waveforms are prestored in both the encoder and the decoder. The encoder tries each of the possible excitation sequences and determines which has the lowest error. It then sends the number describing the position of this codeword in the code book along with a power level. Typically the code book index is represented with about 10 bits (to give a code book size of 1,024 entries) and the gain is coded with about 5 bits. Thus, the bit rate necessary to transmit the excitation information is greatly reduced—around 15 bits compared to the 47 bits used, for example, in the GSM RPE codec. Considerable research is underway to determine means to reduce the number of excitation sequences that need to be tried for each sample of speech so as to simplify the design of CELP coders. For example, it has been found that the CELP codec structure can be improved and used at rates below 4.8 kbps by classifying speech segments into one of a number of types (e.g., voiced, unvoiced, and transition frames). The different speech segment types are then coded differently with a specially designed encoder for each type.

The GSM full-rate speech codec is an example of a RPE codec operating at 13 kbps. The input speech is split up into 20-ms-long frames, and for each frame a set of eight short-term predictor coefficients are found. Each frame is then further split into four 5-ms subframes, and for each subframe the encoder finds a delay and a gain for the LTP. Finally, the residual signal after both short- and long-term filtering is quantified for each subframe. This residual signal is decimated into three possible excitation sequences, each 13 samples long. The sequence with the highest energy is chosen as the best representation of the excitation sequence, and each pulse in the sequence has its amplitude quantified with three bits. At the decoder the reconstructed excitation signal is fed through the long term and then the short-term synthesis filters to give the reconstructed speech. A postfilter is used to improve the perceptual quality of this reconstructed speech.

An area of recent progress has been in variable-rate speech coders. Such coders generate more information when the user is speaking and less information when he or she is not. Advanced variable rate coders typically have around eight different rates that are selected depending on the amount of information required to characterize the speech. A 20-ms section of speech is sampled, the amount of voice activity measured, and an appropriate coder rate set for that section. The first of this type of coder was developed by Qualcomm and called PureVoice™. More recently the GSM standards committees are working on an *advanced multirate coder* (AMR coder). Such speech coders are extremely complex and beyond the scope of this book; suitable references for further reading would be [1, 6]. Variable rate coders have the advantage that less interference is generated when the user is generating less information; the reason why this adds more capacity is detailed in Section 9.4.

The rate at which codecs will improve in future years is far from clear. At the time that GSM standardization was taking place in the late 1980s, 13-kbps voice coders were available and the view was that soon 6.5-kbps voice coders of comparable quality would become available. By the late 1990s, such coders were still not available and the trend of rapid reduction in coding rate seemed to have come to a halt. However, there is still, in principle, much reduction in coding rates that can be achieved, and despite the current slowdown in progress, it seems likely that further advances in speech coding will be found.

2.4.3 Secure transmission

Most analog cellular systems had little in the way of encryption. Anybody with a scanner able to scan the cellular frequency bands was typically able to pick up a cellular conversation. Although not initially a problem, eventually there were a number of scandals resulting from politicians having their calls intercepted, culminating in an intercepted phone call from Princess Diana, the so-called "squidgy" episode. Analog phone systems also started to experience fraud problems. The earliest of these related to cloning. Fraudsters with scanners would wait in places where people were likely to turn on their phones, such as airports. When the phone was turned on it sent its identity to the base station. The fraudster intercepted this message and placed the identity of the user's phone into their own phone. All the calls the fraudster made were now billed to the original user, who would not discover this until his or her bill arrived the following month. This cloning problem reached epidemic proportions in some countries. It was obvious to the designers of digital phone systems that more advanced security was required.

Digital phone systems have security algorithms in place to overcome the following problems:

▶ Encryption, which overcomes eavesdropping of a phone call by an unauthorized listener;

▶ Authentication, which ensures that a mobile cannot be cloned;

- Temporary identities (see Section 2.4.3), which ensure that eavesdroppers cannot determine the approximate location of a subscriber;

- Some systems such as TETRA also have a mechanism whereby the phone can authenticate the network to ensure that a "dummy" network has not been set up that might prevent the user logging onto the valid network.

All security mechanisms follow the same basic principle. At the time of manufacture the mobile (actually the *Subscriber Identity Module* (SIM) card in GSM) is given a unique and secret number. This number is only known to the operator—it cannot be accessed by the user. The secrecy then resides in keeping this number confidential. When the network wishes to authenticate a user, it sends him or her a random number. It sends the same random number to the *home location register* (HLR; see Section 3.2), where the secret number is also stored. The HLR combines the random number with the secret number in a "one-way" algorithm that produces a particular result. A one-way algorithm is one where it is relatively simple to calculate the result given the two inputs but almost impossible to calculate one of the inputs given the other input and the output. The result is sent to the MSC. The mobile sends the random number to the SIM, which also returns the result. The result is transmitted back to the *mobile switching center* (MSC), which compares the two and, if they agree, authenticates the mobile. Because the random number will be different each time, even if the exchange of random number and result is overheard, this is of no use to the would-be cloner. Note that the random number itself is never sent anywhere, even down wires. This prevents any possibility of interception.

The mobile and the HLR also pass the random number and confidential number into a different algorithm that is responsible for producing a mask that is typically a binary string some 64 bits long. The transmitter then XORs[3] the user data by this mask in order to derive the transmitted information. This process is shown in Figure 2.19.

The receiver XORs the received data with the same mask. This results in the original data being decoded. Since the covert listener does not know the mask, he or she is unable to decode the data and hence cannot

---

3. The XOR, or "exclusive OR," process is one whereby the output is "1" if either, but not both, of the inputs are "1."

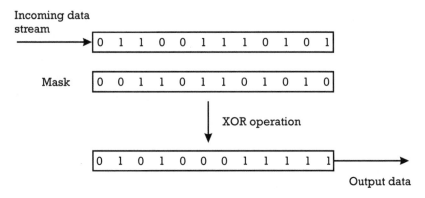

Figure 2.19   Multiplication by a mask.

listen to the conversation. However, even this is not enough. If the mask remains unchanged for a long enough period of time, it might be possible to use statistical methods to break the mask. Hence, the mask is derived from the confidential number, random number, and the frame number, which increases in GSM each frame (see Section 3.2), around every 6 sec. As a result, the mask changes sufficiently frequently that statistical methods will have insufficient information to break the code.

The one-way algorithm   Most one-way algorithms work on the basis of modular arithmetic. The secret code is raised to the power of the random number, modulo N. So, for example, if the data byte of the secret code were 2, the random number for that byte 3, and $N = 7$, then $2^3 = 8$ and 8 Mod 7 = 1. In order to determine the secret code, an eavesdropper would need to raise the encrypted information to the power of a decoding code also modulo N. Although the eavesdropper may know the transmitted result and the random number, calculating the decoding code for any particular random number is almost intractable if the random number is sufficiently large. In summary, it is the modular arithmetic that gives the function its one-way nature since information is "lost" during the modular operation, leaving the eavesdropper without the capability to work backward through the function.

2.4.4   Overcoming channel imperfections

Because of the fading environment explained in Section 2.2 and the presence of interference, it is rare that a mobile radio transmission is without errors. Most radio systems have error rates of around 10^{-2}, or 1 in 100 bits in error. The simplest way to reduce this error rate would be to increase the transmitted power such that even in a deep fade there would still be sufficient signal strength to reliably receive the signal. However, this would just increase the interference to the next cell, resulting in an increased error rate in that cell. An alternative, and better, approach is to add redundancy to the transmitted signal. This has the effect of increasing the bandwidth required for transmission but allowing the receiver to use knowledge of the redundancy to remove errors. This tradeoff of bandwidth for decreased error rate is the basis of error correction systems. The best error correction systems are those that provide the greatest reduction in error rate for the smallest increase in bandwidth. In fact, Shannon's theory (see Section 2.6) relates the bandwidth and the signal to noise ratio and can be used to judge how well error correction systems measure against the ideal.

There are two different types of error control systems, those based on block coding and those based on convolutional coding [1]. Both work by adding extra information to the data to be transmitted and then using a knowledge of the redundancy in order to correct errors in the original data. The difference between block and convolutional coding is in the manner in which the redundancy is added. Block codes add a block of extra data after the information to be transmitted. Convolutional codes modify the data itself, adding redundancy in the process.

A typical block code uses a matrix known to the transmitter and the receiver. The input data is placed in an empty matrix and multiplied by the agreed matrix. The transmitter then sends the input information and the result of the matrix multiplication. At the receiver, the received information is placed in an empty matrix and the same multiplication performed. The receiver checks for differences between the transmitted result and the derived result and uses these differences to compute where the errors are most likely to have occurred. Convolutional codes are examined in more detail in the next section.

2.4.4.1 Encoding

A diagram of a highly simplified convolutional coder is provided in Figure 2.20.

The manner in which this encoder works is as follows. The input datastream is passed into the shift register, with the data being shifted to

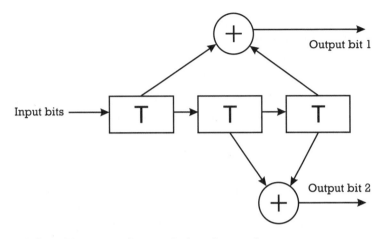

Figure 2.20 Diagram of convolutional encoder.

the right each time a new input bit is presented. For each time the input is shifted, two output bits are derived based on the addition (modulo 2) of the first and third and of the second and third bits. Redundancy is added into the signal because if the first three inputs had been 111, then these would be shifted to the right as a new bit arrived, resulting in the encoder containing either 111 or 011, depending on whether the new bit was a 1 or a 0, respectively. In the case of a 111 input, the output would be 00; while in the case of a 011 input, the output would be 10. The receiver could then deduce that if 10 or 11 had been received that an error would have occurred during the transmission process. By piecing together the most likely input stream given a particular received waveform and taking account of the fact that each pair of bits is related to the previous pair, the receiver can decode a certain number of errors. The optimum way to do this is by using a Viterbi decoder.[4] Such decoders are widely used in mobile radio for error correction, equalization, and reception, and the concepts of the Viterbi principle and optimal receivers are important for wireless professionals to grasp. Viterbi decoders are discussed at the end of this section.

Errors in mobile radio systems tend to arrive in blocks when the mobile passes through a fade. However, a burst of errors can overwhelm an error correction system that can only correct a certain number of

---

4. As an aside, the Viterbi decoder is named after Andrew Viterbi, one of the cofounders of Qualcomm, a mobile radio equipment manufacturer.

errors in a block of data. This problem is overcome by randomizing the errors using a device known as an interleaver. This essentially mixes up the bits before transmission and reorders them on reception. A sequence of bits in error, when reassembled correctly at the receiver, would then be more randomly distributed. There are many different ways to mix up data, some of which have slight advantages over the others. The most simple to imagine is the block interleaver, shown in Figure 2.21, where data are fed into the rows in the left-hand matrix until the matrix is full. They are then fed out in columns to be transmitted. Clearly, they need to be inserted into the columns in the receiver and the receiver matrix needs to be full before the rows can be read out, with the data restored to their original order. Should a burst of errors have occurred during transmission, then all the bits in one column might have been in error. However, when reassembled, this will result in one error in each row, a more even distribution of errors than would otherwise have been the case.

Unfortunately, interleavers introduce delay in the transmission process because no information can be transmitted until the matrix at the transmitter is full. Similarly, no information can be read out of the receiver until the matrix at the receiver is full. In the case of Figure 2.21, the delay would be equivalent to 72 bit periods because 36 bits need to be read into the transmitter interleaver before any can be transmitted, and the receiver must wait for all 36 bits to be transmitted before any decoding can take place. Adding delay to the speech makes the system sound increasingly like a satellite link, with the difficulties that humans have in communicating when there is a long delay, and clearly should be avoided. There is a difficult balance to be achieved between an interleaver size (known as the interleaving depth) and the delay that results in the best

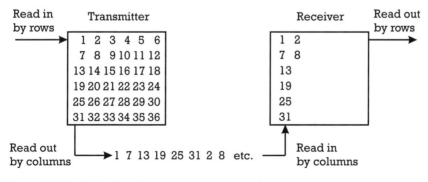

Figure 2.21 Interleaving.

overall speech quality, balancing the adverse effects of delay against the adverse effects of uncorrected errors.

2.4.4.2 Decoding

If data are convolutionally encoded using an encoder of the form shown in Figure 2.20 but with a more practical longer shift register, then the bit sent at any particular point in time is dependent on the bits already in the shift register. Thus, to decode a particular bit, it is important to have correctly received the previously transmitted bits so that the contents of the shift register can be reconstructed and the input data deduced. But the previously transmitted bits themselves depend on the bits transmitted previous to that, and so on. Hence, it might be deduced that the optimal receiver would store the entire received waveform and then find the most likely transmitted sequence that fits the received waveform. To do this, the optimal receiver would locally generate every possible sequence of 1s and 0s that might have been generated by the user, pass them through a copy of the convolutional encoder used by the transmitter, and compare the resulting datastream with that received. The sequence of 1s and 0s that most closely matched the received sequence would then have been deduced to have been transmitted, with any differences between the received waveform and the best-fit reconstructed waveform being transmission errors. The input data corresponding to this sequence can then be output from the receiver as the corrected received data.

The problem with such a decoder is its complexity. For an input message length of only 20 bits, the number of possible combinations of messages is over 1 million (2^{20}), and the complexity rises exponentially with increasing message length. Such an approach is clearly impractical. Viterbi studied this impractical receiver and found that, in practice, it was not necessary to examine every possible input sequence. To understand the Viterbi decoder, the encoder shown in Figure 2.20 is used as a very simple example.

It is normal to represent the receiver as a trellis of possible states, as is shown in Figure 2.22.

Here the possible states of the encoder (i.e., the bits that could be in the shift register in the encoder) are down the left-hand side. Because there are three bits in the shift register, there are eight possible states. From any given state, it is only possible to go to one of two other states, depending on whether the new input bit is a 0 or a 1, as shown

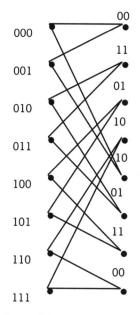

Figure 2.22 The start of a trellis diagram.

in Figure 2.22. In this figure, the transmitted data bits (bit 1 and bit 2 in the encoder diagram) are also shown on the right of the trellis.

Now imagine the coder always starts in the 000 state and look at a number of input bits. The trellis now starts to look like that shown in Figure 2.23.

In this figure, the possible paths from a particular starting state (000) are shown in bold. Now imagine that each of these possible paths was given a score depending on how close it was to the received data. For example, in between step 1 and step 2 there are two possible paths, the first (state 000 to state 000) relating to transmitted data 00 and the second relating to transmitted data 10. If the received data were 00, then for the first path the difference between the nominal data and the received data would be 0 (they are both the same), whereas for the second path the difference would be 1 (because one of the two bits was different). These scores are remembered and are attached to the nodes, so the node 000 on step 2 has a score of 0 and the node 100 has a score of 1. This process continues between each step, with the score on step 3 equal to the score on step 2 plus the difference between the received data and the nominal data for the transition. So, for example, consider the state 100 on step 2.

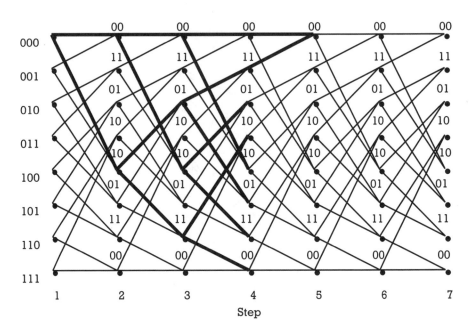

Figure 2.23 The trellis becoming more complete.

Transitions to states 010 and 110 are allowed relating to transmission of 01 and 11, respectively. Say the received data were 00, then node 010 would have a score of 2 (because node 100 had a score of 1 and the difference is 1) and node 110 would have a score of 3 (difference of 2).

A key point is reached at step 5. Here, two paths converge into each state (only those converging into state 000 have been shown for clarity). Viterbi said that whichever paths are followed into the future from state 000, they will always select as their tail the path entering state 000 with the lowest score. Hence, whenever two paths converge, the scores for each path are calculated and the path (or paths in the case of more complex encoders) with the highest scores are discarded. So for each node in the trellis, it is only necessary to remember the score at that point and the incoming path relating to that score. Viterbi found another key point. If the highest score paths were discarded, then at any particular point, if the step in the trellis say 20 steps prior to that currently being processed was examined, all the current valid paths at the current step would typically pass through only one state this far back in the trellis. Viterbi found from experimentation that this point was around 7 times the constraint length in steps back from the current point, where the constraint length broadly relates to the number of bits in the shift register (e.g., three in the case of this example).

These discoveries dramatically simplified this optimum form of decoding. Instead of having to trace millions of paths, only 2^c, where c is the constraint length, paths need decoding at any point, and only $2^c \times 7c$ pieces of data need to be remembered (168 in the case of this example). This is still not trivial, but a massive improvement on the previous optimal detection. Remember that the Viterbi approach is still optimal in that it compares the received data with every possible piece of transmitted data over a large number of bits in order to deduce the most likely received output.

The form of Viterbi decoding just described is often called "hard decoding" because the received data are quantified to a 1 or a 0 prior to the Viterbi decoding process. This is not essential; unquantified received items of data such as 0.83 could be input into the decoder. This increases the complexity slightly because more memory is required to remember the noninteger scores, but better performance is typically achieved as a result.

The decision of whether to use a Viterbi decoder is always difficult. It is an optimal decoder and so by definition generates the best results. However, it also has much greater complexity than nonoptimal decoders. The decision typically depends on the number of states in the encoder. The complexity of the Viterbi decoder rises exponentially with the number of input states whereas the complexity of most other decoders rises only linearly. Hence, as the encoder becomes increasingly large, the Viterbi decoder becomes less likely to be selected due to its complexity.

2.4.5 Frequency and phase modulation

After having undergone the process described in the previous sections, there is an encoded and error-corrected digital data stream consisting of square-wave binary pulses. If such a waveform were transmitted directly, then it would use vast amounts of spectrum. Fourier transform theory shows that an instantaneous change in the time domain (e.g., the rising edge of a binary pulse) requires an infinite amount of bandwidth in the frequency domain. Something more needs to be performed before the signal can be transmitted, and this is the process of modulation and pulse shaping.

Modulation is typically divided into three classes:

▶ *Phase modulation* (PM), where the phase of the carrier waveform is changed depending on the information to be transmitted;

▶ *Amplitude modulation* (AM), where the amplitude of the carrier waveform is changed depending on the information to be transmitted;

▶ *Frequency modulation* (FM), where the frequency of the carrier waveform is changed depending on the information to be transmitted.

Frequency and phase are closely related: an increase in the frequency of a wave will result in a change in the phase for a receiver whose reference frequency does not change. Similarly, a change in the phase will result in a short-term change in the frequency. Because of this close relationship, frequency and phase modulation can be considered as part of the same class of modulation waveforms. Examples of waveforms for phase and amplitude modulation when an analog signal is transmitted are shown in Figure 2.24. Typically, phase modulation is less susceptible to interference and is more widely used.

In the case of digital transmissions, the modulated waveforms are shown in Figure 2.25.

Here it can be seen that there is an abrupt transition between the different states as the modulating data change. Such abrupt transitions cause the transmitted spectrum to become much broader, increasing the interference to other users.

Improvements can be made in the use of phase modulation if *minimum shift keying* (MSK) is adopted. In MSK the phase of the signal is advanced or retarded by 90 degrees over the symbol period, typically in a linear manner. This avoids the sharp changes in phase shown in Figure 2.25. A more advanced category of phase modulation is *Gaussian MSK* (GMSK), where instead of the phase varying in a linear fashion over the symbol period, the rate of change is limited with a Gaussian response. The bandwidth of the transmitted waveform for simple *binary phase shift keying* (BPSK) is shown in Figure 2.25(d); MSK and GMSK are shown in Figure 2.26.

In this figure, the stylized and simplified transmitted power is shown at the carrier frequency (represented as 0 on the x-axis) and at increasing frequency deviations from the carrier, where the bandwidth of the

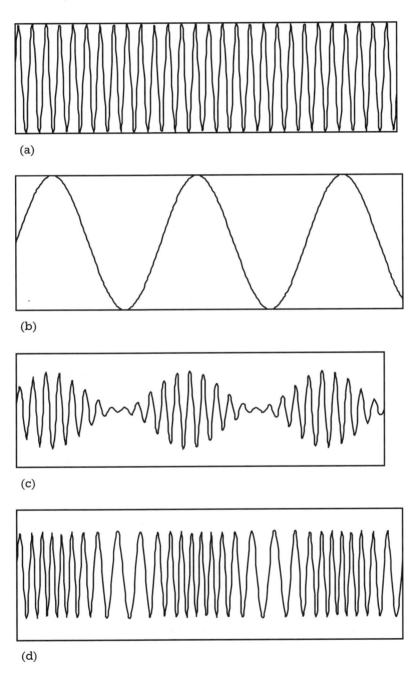

Figure 2.24 (a) Carrier waveform, (b) modulating waveform, (c) amplitude modulated waveform, and (d) frequency modulated waveform.

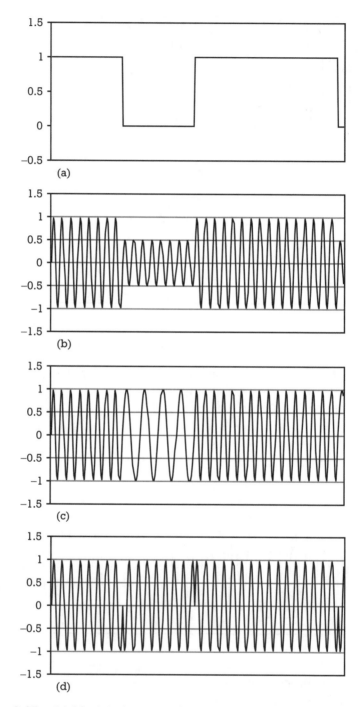

Figure 2.25 (a) Modulating waveform, (b) amplitude modulation, (c) frequency modulation, and (d) phase modulation.

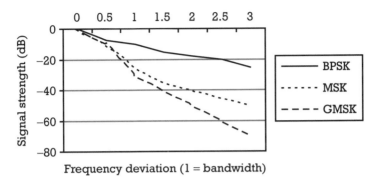

Figure 2.26 The comparative performance of BPSK, MSK, and GMSK.

baseband signal is 1. An ideal modulation scheme would result in the signal falling below −100 dB at a frequency deviation of 1. From Figure 2.26, the very poor performance of BPSK is clear with the signal only falling to −20 dB around 2.5 times the bandwidth from the carrier. MSK is significantly better with GMSK, providing slight gains over MSK.

Pulse shaping   Another way of looking at approaches to reducing the transmitted bandwidth such as GMSK is as a filtering process in which the input pulses are passed through a filter before being modulated with a basic PM technique. This is termed pulse shaping. The square pulses are passed through a filter, resulting in rounded pulses. The more severe the filter, the lower the bandwidth of the transmitted signal, but the more the difficulty in decoding the signal. Depending on the bandwidth scarcity, such tradeoffs need to be made by the system designer.

A set of widely utilized filters is the so-called *raised-cosine* (RC) filters, which modify the ideal square low-pass filter response by placing a quarter period of a cosine-shaped curve on both the vertical transitions, as can be seen in Figure 2.27. The bandwidth is controlled by the so-called roll-off factor, defined as the ratio of excess bandwidth above the perfect filter. The RC characteristic is then defined as

$$G(f) = T \text{ for } 0 \le |f| \le \frac{1-\alpha}{2T}$$

$$G(f) = \frac{T}{2}\left(1 - \sin\left[\frac{\pi T}{\alpha}\left(f - \frac{1}{2T}\right)\right]\right) \text{ for } \frac{1-\alpha}{2T} < |f| \le \frac{1+\alpha}{2T} \tag{2.8}$$

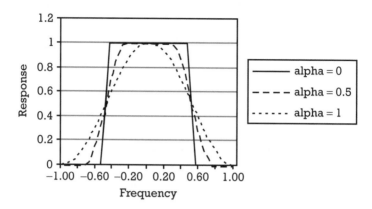

Figure 2.27 Filter response for a range of roll-off factors.

where α is the roll-off factor that controls the severity of the filter and $T = 1/2f$ is a measure of the bandwidth of the signal to be transmitted. Figure 2.27 shows the filter response for a range of different roll-off factors.

An optimal filter would be one with $\alpha = 0$, where there would be no interference into adjacent channels. However, such a filter is impossible to achieve and would result in the input pulses, when viewed in the time domain, being spread over an infinite period. Coupled with each of the frequency responses shown in Figure 2.27 is a corresponding time domain, or impulse response, that shows how the input pulses are spread over time. The higher the value of α, the less the pulses are spread in the time domain.

The impulse response of the RC characteristic plays an important role in deciding upon the choice of in a particular system. Namely, when the channel becomes nonideal and, hence, the ISI is nonzero, the decay of the impulse response must be as rapid as possible to minimize the duration over which a previous transmitted symbol can influence the received signal. In summary, lower roll-off factors yield spectral compactness, but they are also more vulnerable to ISI via nonideally equalized channels. This makes clock recovery much more difficult to achieve.

It is possible to extend the impulse response of the filter over more than the duration of one symbol period. This allows a compact frequency response that does not need large guard bands between radio channels but "smears" the information in one pulse into the next pulse. Such a technique is called partial response signaling. The effect for the receiver is identical to ISI and requires the use of an equalizer, as discussed in Section 2.4.4, to recover the original data. In modern radio systems, added complexity is a small price to pay for increased spectrum efficiency and, hence, systems such as GSM spread input pulses over three symbol periods to reduce the transmitted bandwidth. Such partial response filters are often categorized by their *bandwidth-time* (BT) product, where the product is given by 1 divided by the number of symbols across which the data are spread. A system with a BT product of 1 only spreads the input pulses over one symbol period, whereas a system with a BT product of 0.33 spreads the input pulses over 3 symbol periods. GSM uses a BT product of 0.3. One of the most normal uses of partial response filters is with GMSK, where the Gaussian response is modified according to the required BT value. Figure 2.28 shows the performance of GMSK with a range of different BT values.

The gains in performance from using a smaller value of BT are clear. However, while a system with a BT = 1 would not require an equalizer to decode the modulation, one with BT = 0.5 would require an equalizer operating over 2 symbol periods and one with BT = 0.3 over 3.3 symbol periods (in practice, this would mean over 4 symbol periods). If the optimal Viterbi equalizer is used, the relative complexity of the three schemes would be 1, 4, and 16 for BT = 1, 0.5, and 0.3, respectively.

Quadrature modulation   In order to encode more bits of information into the same radio spectrum, both the amplitude and the phase can be changed simultaneously, with some bits of the input information being encoded onto the amplitude and others onto the phase. This is

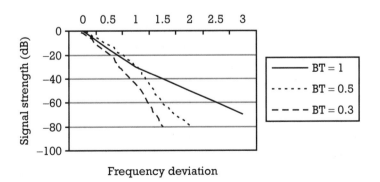

Figure 2.28   Comparative performance of GMSK with different BT values.

termed *quadrature amplitude modulation* (QAM) because the amplitude is being varied on both the in-phase and quadrature arms of the modulating waveform. Broadly, the advantage of QAM is that more information is encoded for the same spectrum. The disadvantage is that a greater *signal-to-noise ratio* (SNR) is required. On twisted pair cables, a high SNR is available and so QAM is a natural choice; whereas with mobile radio, a low SNR is typically experienced and hence QAM cannot be simply deployed.

A typical QAM constellation is shown in Figure 2.29, where the I- and Q-axes represent the in-phase and quadrature modulation of the carrier, respectively. The in-phase wave is that modulated using a sine wave, while the quadrature wave is that modulated using a cosine wave. Both of these waves can be modulated onto the same carrier simultaneously because they are orthogonal; that is, the presence of one does not affect the data carried on the other. A particular point on the constellation of Figure 2.29 is transmitted via suitable amplitudes of the in-phase and quadrature waves. The result can also be considered to be a vector or phasor with a particular amplitude (the length of the vector) and a particular angle (the angle of the vector relative to the I-axis). Both representations are interchangeable.

The apparent advantage of QAM is that, having a constellation with more points than the two used in binary in-phase modulation (known as BPSK) or the four used in quadrature single amplitude modulation (known as QPSK) as shown in Figure 2.30, more bits of information can be transmitted on each symbol. A symbol is the information transmitted during one modulation period; this information can include more than one bit of information in the case that there are more than two potential modulation states. In the case of Figure 2.29, four bits per symbol can be transmitted compared to one for BPSK and two for QPSK (hence the reason for the term *ISI* rather than *interbit interference because* there can be more than one bit per symbol). It might seem that QAM was advantageous because of this capability to carry more information. However, this is not true in mobile radio. Because the points in a QAM constellation are closer together than in a QPSK constellation, the signal needs to be received with less noise in order to be correctly decoded. In dense mobile radio networks, most of the noise is cochannel interference, generated from nearby cells using the same frequency. The net result is that QAM systems need to work with a larger cluster size than QPSK systems. The increase in cluster size more than offsets the gain in the number of bits per symbol, with the result that QAM systems tend to have a lower overall

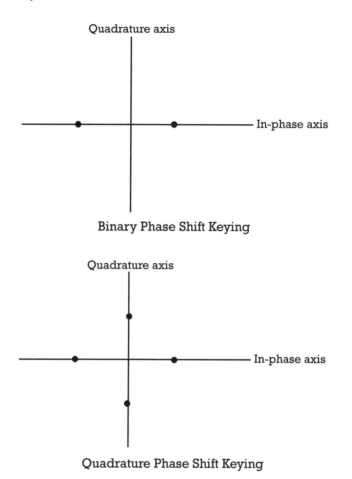

Binary Phase Shift Keying

Quadrature Phase Shift Keying

Figure 2.29 BPSK and QPSK constellations.

spectrum efficiency. This was proved rigorously in [7] and, hence, it seems unlikely that QAM will be widely used for cellular. Even in the world of WLL, where cochannel interference is less severe as a result of the directional antennas used, QAM has been shown not to be advantageous [4]. Only in fixed communications, where there is no cochannel interference, is QAM suitable. Those interested in finding out more about this form of modulation should refer to [8].

Other types of modulation   There are many other variants of modulation that are typically used with quadrature transmission to overcome the problems caused when a transition from one symbol to another

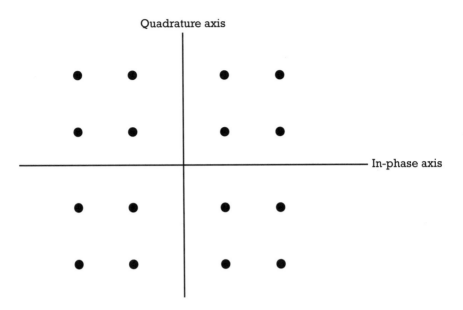

Figure 2.30   A typical QAM constellation.

causes the modulating signal to cross the origin of the constellation diagram. This would require the transmitted signal to fall momentarily to zero, resulting in full amplitude modulation of the carrier. If the amplifiers used are not linear, the transition through zero is likely to result in distortion and a broadening of the transmitted spectrum. Techniques to overcome this problem include $\pi/4$ QPSK and *offset QPSK* (O-QPSK). In $\pi/4$ QPSK the constellation points "rotate" by 45 degrees each symbol period, preventing any transition from passing directly through the origin. In O-QPSK the transition in the Q-axis is delayed by half a symbol period to the transition in the I-axis, again preventing a transition through the origin. These forms of modulation are discussed in more detail in [8].

2.4.6   Clock recovery

It is important that both the transmitter and the receiver are synchronized because otherwise the receiver does not know when to sample the incoming waveform nor with what to compare any change in phase of the incoming waveform. Receivers may require both clock and carrier recovery systems that both attempt to derive information about timing from the received signal, often in a similar manner. While carrier recovery is only necessary in coherent demodulation systems (which do not use differential coding), clock recovery is required in all schemes, and accurate clock recovery is essential for good data transmission. Confusion often exists between clock and carrier recovery. Clock recovery attempts to synchronize the receiver clock with the baseband symbol rate transmitter clock, whereas carrier recovery endeavors to align the receiver local oscillator with the transmitted carrier. In this section a brief review of some of the more widely used clock recovery schemes is provided while carrier recovery schemes are considered in the next section. More details can be found in [8].

There exist a number of different clock recovery techniques. However, many apparently different clock recovery techniques can be shown to be equivalent, and this is not surprising since they must all make use of the properties of the received waveform. Basic clock recovery systems include times-two, early-late, zero-crossing, and synchronizer clock recovery systems.

Times-two clock recovery   The most fundamental of all clock recovery schemes is the times-two, or squaring system. If the received demodulated signal is squared, then it will possess a periodic component at the symbol rate. A bandpass filter tuned close to the symbol rate will extract this periodic signal, allowing derivation of the required timing information. The times-two clock recovery works best for binary modulation schemes, but not so well for multilevel ones. The reason for this is that the multilevel scheme has a reduced component at the symbol frequency due to the increased possibility of non-zero-crossing transitions.

Early-late clock recovery   Another well-known form of clock recovery is early-late clock recovery. While times-two clock recovery exploits the whole of the incoming waveform, early-late clock recovery works on the peaks in the received waveform. The basic assumptions made by the early-late method are that the peaks in the incoming waveform are at the correct sample point and that these peaks are symmetrical. This is often true, but for some modulation schemes, such as partial response RC, neither of these assumptions is valid. It may still be possible to use early-late clock recovery with these modulation schemes, but the timing jitter will be increased. The early-late scheme first squares the incoming signal to make all peaks positive and then takes two samples of the received waveform, both equispaced around the predicted sample point. If the predicted sample point is aligned with the correct sample point and the previous assumptions are correct, then the sample taken just prior to the sample point—the early sample—will be identical to the sample taken just after the sample point—the late sample. If the early sample is larger than the late sample, this indicates that the recovered clock is sampling too late, and if the early sample is smaller than the late sample, this indicates that the recovered clock is sampling too early. It is normal to filter the difference of each pair of samples to reduce the effect of random noise on the system.

Zero-crossing clock recovery   Similar to early-late clock recovery in that it looks for a specific feature in the received waveform, is zero-crossing clock recovery. This works on the premise that with symmetrical pulses, the received waveform will pass through zero exactly midway between the sample points. The receiver detects a change in the polarity of the received signal and, if this does not occur midway between predicted sample points, speeds up or slows down its reconstructed clock appropriately.

Synchronizer clock recovery   A completely different clock recovery system is the synchronizer, which is the scheme used by GSM. In this system the transmitter periodically sends a sounding sequence, for which the receiver searches by performing an auto-correlation at a rate significantly faster than the symbol rate (normally four times faster). The oversample point at which the maximum correlation occurs is taken to be the correct sample point and is assumed to remain correct until the next sounding sequence is received. This scheme is simple but requires a sounding sequence to be periodically inserted into the datastream, increasing the bandwidth requirements. It also requires that the transmitter and receiver clocks do not drift significantly over a block, which may place constraints on the circuitry that can be employed. Synchronization has the advantage that it performs well regardless of the modulation scheme in use.

2.4.7   Carrier recovery

Carrier recovery is required when coherent detection is used. Coherent detection requires the receiver to determine the absolute phase and amplitude transmitted. The alternative to coherent detection is to use differential coding. In this case, the transmitted information is encoded as the difference between the previous symbol and the current symbol. The receiver can decode this information by comparing the previous and current symbols and, hence, does not need to understand the absolute phase or amplitude transmitted. Differential coding is widely used because of the simplification in the receiver architecture but has the disadvantage that two errors are generated when only one error would be generated with coherent detection. These two errors result because if one symbol is received in error, there will be a resulting error when decoding the difference between it and the previous symbol and between the next symbol and this symbol. This additional 3-dB penalty is a serious price to pay for resulting simplicity. For a more detailed discussion of when to adopt coherent and noncoherent detection, see [4].

A common problem for most carrier recovery schemes is an inability to resolve phase ambiguities. So for constellations such as QPSK, phase lock can be established at all multiples of 90 degrees (because it is not possible to determine the phase change imposed by the propagation channel), but without additional transmitted information it is not possible to resolve the angle at which phase lock has been established. This problem can be overcome by sending a sounding sequence or via the use of differential coding. Some of the more common carrier recovery systems are summarized next.

Times-*n* carrier recovery   As with clock recovery, a common form of carrier recovery for binary schemes is the times-two or the squared carrier recovery. In this scheme the incoming carrier wave is squared, leading to a periodic component at the carrier frequency that can be isolated using a filter and phase locked loop. This scheme can only work with binary modulation.

Decision-directed carrier recovery   A carrier recovery system that is completely different from all the carrier and clock recovery systems discussed so far is decision-directed carrier recovery. In fixed link systems this has proved to be one of the most popular carrier recovery schemes. In this scheme, a decision is made as to which was the most likely constellation point transmitted given the received symbol. This is normally the constellation point closest to the received symbol, as shown in Figure 2.31. It is then assumed that the phase difference between the received symbol and the constellation point is due to carrier recovery drift; the receiver carrier recovery is updated accordingly.

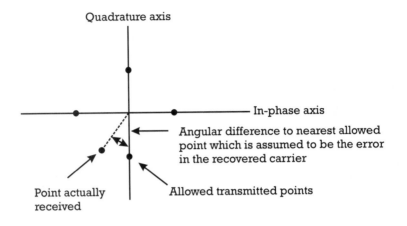

Figure 2.31   Decision-directed carrier recovery.

Decision-directed carrier recovery has the advantage that it can work with all modulation schemes. Decision-directed methods normally have an associated *bit error rate* (BER) threshold. If the BER is less than this threshold, then such methods operate extremely well since they effectively remove noise on the input phasor. If the BER is more than this threshold, then catastrophic failure can result whereby the constellation point selected in the receiver is the wrong one. The update signal to the carrier recovery system is therefore wrong, driving the carrier further from the correct frequency and further increasing the chance of error. Whether decision-directed carrier recovery will be suitable will depend on this BER threshold, which itself is dependent on the modulation scheme and the quality of the radio channel.

2.4.8 Multiple access

Multiple access techniques have become a key topic as a result of the CDMA/TDMA debate discussed in Chapter 14. The selection of multiple access method has some important implications that are discussed in more detail in this section.

There are three key multiple access techniques:

 ▶ *Frequency division multiple access* (FDMA), where the total spectrum assignment is divided into a number of discrete frequencies;

 ▶ *Time division multiple access* (TDMA), where the total spectrum is divided in time between a number of users;

▶ *Code division multiple access* (CDMA), where neither the frequencies or the time are divided but users are distinguished through the use of a special code.

Many other multiple access techniques have been suggested, although most are used in combination with one of the aforementioned three approaches rather than on their own. Worthy of note is *packet reservation multiple access* (PRMA), where the mobile waits until their user is speaking and then reserves a channel by sending a random access message. When the user stops speaking, the channel is returned to the pool. PRMA is efficient in that channels are only used when users are speaking, but there is a risk that a channel is not available when a user starts speaking, resulting in "clipping" of the speech. For this reason, PRMA is not used in any voice-based mobile radio systems. *Space division multiple access* (SDMA) uses directional antennas at the base stations that follow mobiles using beam-forming technology. The narrow antenna beamwidth means the same frequency can be reused in the same sector if the two mobiles are far enough apart. The difficulty with SDMA is the complexity of beam-forming antennas, but these are expected to become feasible in the near future; indeed, there is already a number of manufacturers offering "smart antenna" products.

Each of the three key multiple access methods is now described in more detail.

FDMA   FDMA is the simplest of all access techniques to understand and is the most widespread multiple access technique in use today. The total spectrum allocated to an operator is divided up into a number of equal channels and each user who requires access to a channel is given one of these frequencies for the duration of their call. This is shown symbolically in Figure 2.32. FDMA has been widely deployed because of the simplicity of building FDMA systems. An FDMA mobile needs a transmitter and receiver that can be moved to any given frequency within the assigned band and a filter arrangement that can filter out the unwanted transmissions.

FDMA is a relatively inefficient scheme. It is very difficult to construct a transmitter that keeps all radiated energy within the frequency band assigned, as was shown in the discussion on pulse shaping in Section 2.4.5. The figures in that section showed that substantial energy was transmitted in the frequency bands used by other mobiles, and

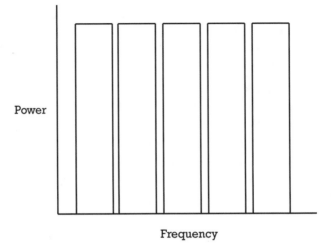

Frequency

Figure 2.32   An ideal FDMA spectrum.

Figure 2.33 shows that this is so even in the relatively advanced GSM system. To overcome this problem, two techniques are adopted. The first is to insert a "guard band" between each frequency to increase the frequency separation between transmissions. The second is to avoid using adjacent frequencies in the same cell. Because, typically, only around 1 in 20 frequencies can be used in any cell with FDMA due to the cluster effect, this latter constraint is not unduly problematic.

A cost disadvantage for FDMA results from the need for each individual radio channel to pass through a separate power amplifier at the base

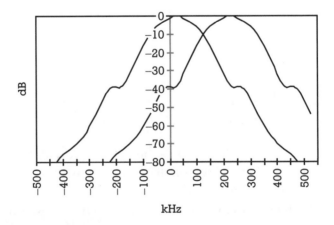

Figure 2.33   GSM spectrum.

station before passing through an expensive high-power combiner and then being transmitted from the antenna. Alternatively, FDMA signals could be combined prior to amplification, but this would require a highly linear amplifier that would be expensive and inefficient in its use of power.[5]

As will be seen, all the access methods have inefficiencies and FDMA is not substantially worse than TDMA. However, the extra cost is significant, and with the widespread deployment of mobile radio systems, FDMA has been used minimally since the early 1990s.

TDMA The rational with TDMA is that if a user is given access to a wider bandwidth than He or she requires, then he or she will only need to transmit for part of the time. During the remainder of the time, other users can transmit their information. To work, TDMA requires precise timing between the transmitter and the receiver so that each user transmits during the time allocated to him or her. It requires a framing structure to separate different users, and it requires that the time between the same user making his or her current and next transmissions is sufficiently small so that unacceptable delay is not added to any speech transmission. The most successful TDMA system is GSM, where each user has access to the channel for $577\,\mu s$ every 4.6 ms, allowing eight users on the same channel.

TDMA also has guard bands between transmissions. In the case of TDMA, the guard bands are gaps in time between one user stopping transmission and the next user starting. These guard bands are required both to accommodate inaccurate timing and to allow the mobile some time to "ramp up" power levels and to ramp them back down again. Such a ramping of power is required since otherwise the instantaneous change from no transmission to full transmission would be akin to sending a square pulse, which, as was explained in Section 2.4.5, results in substantial interference to neighboring frequency bands. The structure of a TDMA burst in GSM is shown in Figure 2.34. A burst consists of a ramp-up period, a longer period of transmission, and a ramp-down period. One mobile ramps up while another ramps down, so the inefficiency is limited to one of the two ramping periods. Hence, around $30\,\mu s$ is "wasted" for each $540\text{-}\mu s$ transmission period.

---

5. Note, however, that such amplifiers are starting to become more common since their use may be required in TDMA systems operating in more than one band, which is likely to drive down their cost over time.

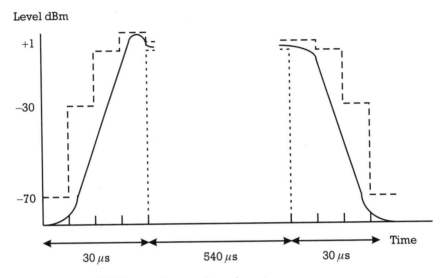

Figure 2.34 GSM ramping up for a burst.

A further problem with TDMA is caused by the fact that the rate of transmission is raised well above that required for FDMA. As explained in Section 2.2, higher transmission rates can result in ISI, which then requires an equalizer to remove the ISI. In turn, this requires the transmission of a sounding sequence, reducing the user data that can be transmitted.

Nevertheless, TDMA is significantly less expensive than FDMA to deploy because it overcomes the need for a separate amplifier for each user. Instead, a separate amplifier is required for each TDMA channel, which can support a number of users. Indeed, the same decoding circuitry, equalizer, and error-correction unit, for example, can all be used on one TDMA channel because only one user can transmit at the same time. It is because of these cost savings that TDMA was selected for GSM.

CDMA   With CDMA, each user transmits across all the allocated spectrum for all the time. In order to fill all the allocated spectrum, they need to increase the amount of information they transmit substantially and they do so by multiplying the user data by an assigned code. It is the redundancy provided by this code that allows the signal from a particular user to be decoded amongst all the interference being generated by the other users transmitting at the same time and on the same frequency. The process of multiplication by a codeword is known as spreading and is shown for an example datastream and codeword in Figures 2.35 and 2.36.

The number of bits in the code assigned to the user will determine the bandwidth of the transmitted signal. In practical CDMA systems, codewords of 64 bits in length are used resulting in the user data taking up 64 times more spectrum than if it were transmitted using FDMA. The length of the codeword is often represented by the letter G. If the number of bits is reduced, then less spectrum is required but less redundancy is

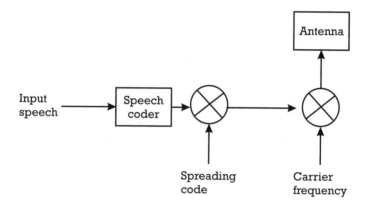

Transmitter—the spreading or coding process

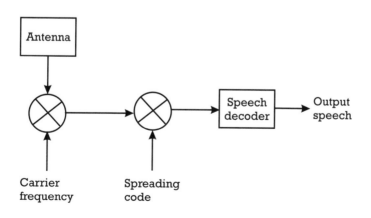

Receiver—the despreading process

Figure 2.35   Process of spreading.

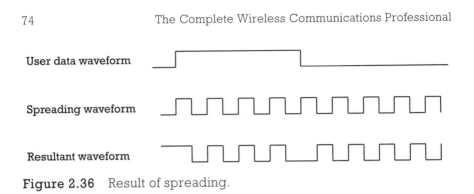

Figure 2.36 Result of spreading.

added. Larger codewords result in a higher transmitted bit rate that brings the advantage of being able to better resolve multipath although at the expense of the operator requiring a larger spectrum assignment. Multipath can be better resolved because, as discussed in Section 2.2, multipath propagation results in fading if the delay on the other paths is less than a symbol duration and results in ISI if it is longer. Fading cannot be mitigated, but ISI can be removed by an equalizer. By using a higher bit rate, any multipath propagation becomes more likely to be ISI rather than fading; hence, the overall propagation is improved, albeit at the expense of requiring a more complex equalizer.

The choice of the code is critically important. The code should have certain *auto-correlation* properties. Auto-correlation is effectively the multiplication of a code by itself when one copy of the code is delayed relative to other copies. Taking a random code of, say, $-1,1,1,-1,1,-1,1,1,-1,1,-1,1$, the process of auto-correlation can be examined. Multiplied by itself with no time shift for the second copy, the result is

First copy	−1	1	1	−1	1	−1	1	1	−1	1	−1	1	
Second copy	−1	1	1	−1	1	−1	1	1	−1	1	−1	1	
Result	1	1	1	1	1	1	1	1	1	1	1	1	=12

Now shifting the second copy by one bit to the right, the result is as follows:

First copy	−1	1	1	−1	1	−1	1	1	−1	1	−1	1	
Second copy	1	−1	1	1	−1	1	−1	1	1	−1	1	−1	
Result	−1	−1	1	−1	−1	−1	−1	1	−1	−1	−1	−1	=−8

This process can be continued for all possible shifts, in this case of ±6 bits from the case where the codes are aligned. The results can be seen in Figure 2.37, where the vertical axis shows the result while the horizontal axis shows the shift from the aligned position.

As can be seen, the largest peak is at zero offset and the result is symmetrical about the origin. Such an autocorrelation would not be suitable for CDMA because some of the results (at, e.g., 5 shift) are almost as great as the result at zero shift. A more suitable autocorrelation curve for CDMA is shown in Figure 2.38, where a large result is obtained only at zero offset.

Many families of codes with such autocorrelation properties are known; for example, Gold codes, named after their discoverer, exhibit such properties.

CDMA signals are recovered by multiplying the received signal, which is the additional of all the signals sent by each of the users, by the code assigned to a particular user. As explained in Section 2.4.3 on encryption, such a double multiplication by the same code results in the original signal being detected. This process within the receiver is known as despreading and results in the wanted signal being enhanced by a factor equal to the codeword length, G. It is this enhancement that raises the level of the signal above the noise generated by the other users and allows it to be correctly decoded. A more detailed example of how despreading occurs can be found in [9].

The process of despreading can only isolate the wanted signal from that transmitted by the other users if they have different codes. Indeed,

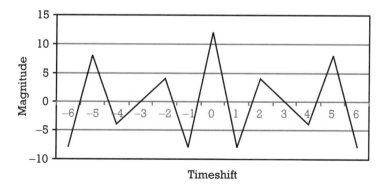

Figure 2.37 The result of autocorrelating –111-11-111-11-11.

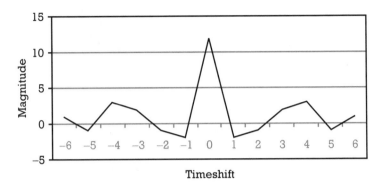

Figure 2.38 A more suitable autocorrelation curve for CDMA.

for CDMA to work perfectly, the codes that they use must all be orthogonal[6] to that used by the wanted transmission. In the case that the codes are orthogonal, when the composite received signal is despread, the signal from the wanted user is enhanced by the factor G, whereas all the other signals are neither enhanced or suppressed. If the codes were different but not orthogonal, then the multiplication process would also enhance some of the data from the other users, making it more difficult to distinguish the data from the wanted user. There are a number of code families that are orthogonal, of which the best known is probably the Walsh family. However, in any one family there can only be as many orthogonal codewords as there are bits in the codeword, that is, G. For more information about codes see [3, 10].

Although in an isolated cell G codes would be sufficient because only G users can be accommodated in any case, the shortage of codes becomes a problem in a network with overlapping cells. If the same frequency is used in each of these cells, then it is quite possible that more than G users could be accommodated across the entire cluster were there enough codes. If only orthogonal codes were employed, the capacity of the total network would be restricted to G/K, where K is the cluster size that would be required to avoid interference between users deploying the same code. In practice, K would probably need to be around 12; hence, with $G = 64$ (a typical value) only five users could be accommodated per cell. This problem is overcome using nearly orthogonal code families such as the

---

6. Codes are said to be orthogonal if when the two codes are multiplied by each other, the result is zero. For more details about the basic mechanism of coding and decoding, see [1].

pseudorandom noise (PN) codes. In such families there is no shortage of codes, but the interference from other users becomes slightly more problematic.

The use of nearly orthogonal codes also raises a further problem. Because the codes are not orthogonal, some component from other users will be decoded during the despreading process. This may not be problematic when the transmissions from all users are received with the same strength, but if the signal strength from another user is significantly greater than that from the wanted user, then the decoded component from the unwanted user could be greater than the decoded component from the wanted user. To overcome this problem, transmissions from all the users in a cell need to be received at nearly the same signal level. (This is not an issue in the downlink direction because all the signals are sent from the same point and so remain at the same relative signal strength.) However, as the mobile moves around the cell, the propagation channel changes rapidly, requiring the mobile to rapidly change its transmitted power levels to keep the received signal at a constant level. This is known as power control.

Power control systems need to be extremely accurate. If a mobile transmits with too much power, it results in excessive interference to all the other mobiles and hence a loss in capacity. If one mobile transmits with twice the power it should have, it can halve the overall system capacity. Maintaining the correct power is difficult since the propagation environment changes rapidly and the mobile needs to adjust to these changes every millisecond. Some simulated results showing how the system capacity is expected to change with increasing error in the power control system in the mobiles are shown in Figure 2.39.

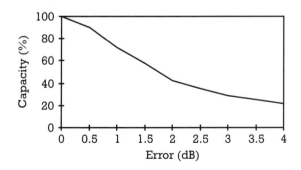

Figure 2.39 Sensitivity to power control error in CDMA systems.

Designing good power control systems is difficult. The systems tend to be combinations of open loop and closed loop systems. The mobile gets its first pass estimate of the path loss to the base station from its received signal. However, because the downlink may be experiencing different fast fading from the uplink, the first pass estimate will only be approximate. The mobile transmits at this appropriate power level based on this first pass estimate and then lets the base station provide closed loop information to achieve accurate power control. The base station sends the mobile periodic signals instructing it to either increase or decrease its power in order to reach the appropriate transmission level. Although necessary to ensure correct system operation, sending such signals is undesirable because they waste radio spectrum. The more frequently the messages are sent, the more accurate the power control will be. Similarly, the more bits in the message, the more steps there can be in power control; but the more messages, the greater the waste of radio spectrum. In practice, simulations have shown that short frequent messages achieve the best results and, hence, most CDMA systems transmit just one bit (indicating either a small increase or decrease in the transmission levels) at a rate of over 1,000 times a second.

In fact, accurate power control would require that the mobile transmit power follow exactly the fast fading environment. As seen in Figure 2.4 this would require extremely rapid changes in power. Such rapid change is typically not possible, and hence, the transmit power control only follows the slow fading and distance related fading. Given that some fast fades are 40-dB deep, the mobile power control could be 40 dB out. Fortunately, in fades the mobile transmits with too little power, resulting in a higher error rate but less interference. However, the fading channel can result in signals around 6 dB higher than average at times and, during this period, the mobile will be generating excess interference. Hence, this is the reason why highly accurate power control is difficult to achieve.

Within the mobile community, there has been much impassioned discussion about whether TDMA or CDMA represents the best access technique. This issue is discussed in more detail in Section 14.4. However, it is becoming increasingly clear that selecting appropriate fixed networks and technologies with economies of scale and widespread deployment is becoming more important than selecting the best access technique.

2.5 Packet and circuit transmission

Transmission of data can be divided into two different types. With circuit-switched transmission, the user requests a transmission resource (such as a TDMA channel) for the whole of the conversation and returns this channel once the conversation has been completed. The transmission resource is known as a circuit. Circuit-switched transmission has the advantage of simplicity, but the disadvantage that even when nothing is being said, nobody else is able to use the channel. It also, typically, takes some time to establish a circuit, which can be disadvantageous when the amount of information to be transmitted is small. This can result in the channel being used for a time period that is the sum of the setup time and the transmission time despite the fact that the transmission time is relatively short.

A solution to these problems is to use packet switching. In a packet-switched system, the user fills up a buffer at the transmitter. When this is full, a resource is requested from the network for the transmission of this single packet. The packet is then dispatched along with address information and is forwarded through the network until it reaches its destination. Because the user does not generate any information when he or she is not speaking, no packets are filled and no transmission resource is required. Further, should the user want to send more information for some reason (perhaps they want to send a data file at the same time as speaking), more packets can be sent, providing greater flexibility.

Packet switching engenders greater complexity. It also has some potential disadvantages for speech and other time-sensitive information. It is quite possible that when the transmitter tries to send a packet the radio resource is temporarily full. The transmitter will need to wait until a resource becomes free. However, if this delay is too great, a pause will result in the speech received at the other end. For this reason, no mobile radio systems to date have transmitted speech using packet transmission. However, as explained in Chapter 6, packet transmission systems for speech have been devised for the fixed network and it seems likely that the increasing importance of the Internet will eventually result in packet transmission over mobile radio networks.

There are two variants of packet switching: connection-oriented and non-connection-oriented. In the case of connection-oriented, a virtual circuit is established between the transmitter and the receiver, passing through the switching nodes, when the first packet is received. All subsequent packets received for the same destination travel via the same route with the result that they will be received in the order in which they are transmitted. In the case of non-connection-oriented, each packet is treated as if no previous packet had been sent. Potentially, a packet could be sent via a different route from the previous packet and, hence, the packets might not arrive at the receiver in the order that they were sent. The receiver then requires a sufficient buffer so that it can correctly order the data prior to presenting it to the user.

Recently, the data traffic on the *public-switched telephone network* (PSTN) has grown to the extent that data volumes are soon expected to exceed voice volumes. In the world of mobile radio, this trend is less pronounced but can be expected to occur eventually. This has lead some to note that modern communications systems should perhaps be packet based rather than circuit based. In the world of fixed networks, a new packet-switching protocol, *asynchronous transfer mode* (ATM) described in Section 6.3, has been designed that can guarantee low delays for high priority traffic, thus allowing it to be used for both voice and data.

As the world changes from sending data over predominately voice networks to sending voice over predominately data networks, the underlying structure of the network will need to change as well. Future radio systems are likely to be increasingly based on packet transmission rather than today's circuit-switched systems.

2.6 Theoretical capacity of mobile radio systems

The maximum capacity that can be achieved from a given piece of radio spectrum was determined by Shannon in the early part of the twentieth century. An understanding of this theory allows the advances that might be achieved in mobile radio systems to be better understood. This section looks at some of the underlying theory in order to determine the maximum capacity that might be achieved within likely spectrum assignments.

The maximum information that can be transmitted per second per hertz of radio spectrum, C, is given by the Shannon law, which states that

$$C = \log_2\left(1 + S/N\right) \tag{2.9}$$

where S/N is the SNR at the receiver. However, in a tightly clustered situation, the performance of the system tends to be dominated by the *signal-to-interference ratio* (SIR) rather than the SNR. Gejji [11] and Webb [7, 12] performed some work in this area. Gejji started with Lee's equation [13], relating the number of radio channels to SIR, and given by

$$m = \frac{B_t}{B_c \sqrt{\frac{2}{3} SIR}} \tag{2.10}$$

where B_t is the total available bandwidth, B_c is the bandwidth required per call, and m is the number of radio channels per cell. He then replaced S/N in Shannon's equation with SIR and substituted for B_c (= R_b/C) within Lee's equation to derive

$$m = 1.224\alpha \frac{B_t \log_2\left(1 + SIR\right)}{R_b \sqrt{SIR}} \tag{2.11}$$

where R_b is the user bit rate and α is a factor relating to the closeness to which the Shannon limit can be approached.

In a well-designed GSM system, engineers typically use a SIR of around 12 dB. For a 1-MHz spectrum assignment and a user bit rate of 13 kbps (the rate of the GSM voice coder) (2.11) predicts a total of 96 voice calls per cell per megahertz. As shown in Chapter 3, GSM achieves much less than this, perhaps 4.5 voice calls per cell per megahertz, while cdmaOne may achieve 12 voice calls per cell per megahertz (these numbers are discussed in more detail in Chapter 14). Hence, Shannon predicts that an eight-fold increase in capacity per cell is still theoretically possible. In some ways, this comparison is unfair. For example, Shannon says nothing about the efficiency of the voice coder. If a mobile radio system with a higher bit rate voice coder that could tolerate more errors were deployed, a smaller cluster size would be possible and the mobile radio system would appear to move closer to the Shannon limit.

In practice, approaching close to the Shannon limit with a mobile radio system is very difficult because of the irregularity of natural features resulting in the need to allow margins when designing networks and also in the nonideal nature of Rayleigh fading requiring complex error control systems and equalizers. Further losses occur in the inefficiencies present in all the multiple access methods as discussed in Section 2.4.8. It is hard to predict how much more closely mobile radio systems might be able to approach this limit in the future, but the complete wireless professional should be wary of systems that claim to approach close to this limit or even go beyond the limit.

References

[1] Steele, R., ed., *Mobile Radio Communications*, London: John Wiley, 1992.

[2] Doble, J., *Introduction to Radio Propagation for Fixed and Mobile Communications*, Norwood, MA: Artech House, 1996.

[3] Siwiak, K., *Radiowave Propagation and Antennas for Personal Communications*, Norwood, AM: Artech House, 1998.

[4] Webb, W., and L. Hanzo, *Modern Quadrature Amplitude Modulation*, London: John Wiley, 1994.

[5] Schwartz M, *Telecommunications Networks: Protocols, Modelling and Analysis*, Reading, MA: Addison-Wesley, 1987.

[6] Redl., S. M., M. K. Weber, and M. W. Oliphant, *GSM and Personal Communications Handbook*, Norwood, MA: Artech House, 1998, Ch. 11.

[7] Webb, W., "Modulation Methods for PCNs," *IEEE Communication Magazine*, Vol. 30, No. 12, Dec. 1992, pp. 90–95.

[8] Franks, L. E., "Carrier and Bit Synchronisation in Data Communications, A Tutorial Review," *IEEE Trans. on Communications*, Vol. COM-28, No. 8, Aug. 1980, pp. 1107–1121.

[9] Glisic, S., *Spread Spectrum CDMA Systems for Wireless Personal Communications*, Norwood, MA: Artech House, 1997.

[10] Prasad, R., *CDMA for Wireless Personal Communications*, Norwood, AM: Artech House, 1996.

[11] Gejji, R. R., "Channel Efficiency in Digital Cellular Communications Systems," *Proc 42nd IEEE Vehicular Tech Conf.*, Denver CO, 1992.

[12] Webb, W., "Spectrum Efficiency of Multilevel Modulation Schemes in Mobile Radio Communications," *IEEE Trans. on Communications*, Vol. 43, No. 8, Aug. 1995, pp. 2344–2349.

[13] Lee, W. C. Y., "Spectrum Efficiency in Cellular," *IEEE Trans. on Vehicular Technology*, Vol. 38, No. 2, May 1989, pp. 69–75.

Further reading

Akaiwa, Y., *Introduction to Digital Mobile Communications*, New York: John Wiley, 1997.

Balston, D., and R. Macario, *Cellular Radio Systems*, Norwood, MA: Artech House, 1994.

Calhoun, G., *Digital Cellular Radio*, Norwood, MA: Artech House, 1988.

Faruque, S., *Cellular Mobile Systems Engineering*, Norwood, MA: Artech House, 1996.

Freeman, R., *Radio System Design for Telecommunications*, New York: John Wiley, 1987.

Meyr, H., *Digital Communication Receivers: Synchronization, Channel Estimation, and Signal Processing*, New York: John Wiley, 1997.

Pahlavan, K., and A. Levesque, *Wireless Information Networks*, New York: John Wiley, 1995.

Webb. W., *Understanding Cellular Radio*, Norwood, MA: Artech House, 1998.

3

Contents

Cellular radio technologies

The essence of success is that it is never necessary to think of a new idea oneself. It is far better to wait until someone else does it, and then copy him in every detail, except his mistakes.

Aubrey Menen

3.1 The range of cellular systems

Many mobile radio professionals will be involved in the world of cellular radio, either on the design and manufacture of hardware or on the deployment and running of cellular networks. This chapter looks at the different technologies in overview. For almost all of these technologies there are specialized books detailing all aspects of the technology [1–5]. Here, the salient differences and key references are introduced.

The wide range of different cellular standards in the world today is shown in

Table 3.1. Broadly these are divided between analog and digital systems and between different regions of the world.

It is worth considering why there are so many different systems. Some countries encourage the adoption of a single system on the basis that it will bring benefits to consumers roaming throughout the country.

Table 3.1
Cellular Systems Worldwide

Name	Analog or Digital	Major Countries/ Regions of Use	Date of Introduction	Frequency Band	Key Features
NMT	Analog	Scandinavia, East Europe, South East Asia	early 1980s	450 MHz and 900 MHz	Allowed roaming between countries
AMPS	Analog	United States and South America	mid 1980s	800 MHz	—
TACS	Analog	West and South Europe, Asia	mid 1980s	900 MHz	Derived from the AMPS system
GSM900	Digital	Worldwide	1992	900 MHz	See Section 3.2
GSM1800	Digital	Worldwide	1994	1,800 MHz	Same as GSM but at a higher frequency
GSM1900	Digital	United States, Chile	1996	1,900 MHz	Same as GSM but at a higher frequency
D-AMPS	Digital	United States	mid 1990s	800 MHz and 1,900 MHz	—
cdmaOne (previously known as IS-95)	Digital	United States, Asia-Pacific	late 1990s	800 MHz and 1,900 MHz	Only system to use CDMA
PDC	Digital	Japan	mid 1990s	800 MHz	—

The acronyms in this table are:

TACS	Total Access Communication System;
NMT	Nordic Mobile Telephone;
AMPS	Advanced Mobile Phone Service;
GSM900	Global system for mobile communications;
GSM1800	GSM at 1,800 MHz;
GSM1900	GSM at 1,900 MHz;
PDC	Personal digital cellular;
D-AMPS	Digital AMPS.

Others allow any systems to be used on the basis that competition will benefit the consumer. There is a certain national pride in some cases at having a "homegrown" radio system—this is particularly true in the case of Europe and Japan. Finally, some companies feel that they have invented a better radio system and try to sell this system in order to make a profit—Qualcomm is a good example of this.

When comparing the different systems, the key parameters from user and operator points of view are the features and facilities that the systems offer, their capacity for a given spectrum assignment, and their cost. Broadly, all the digital systems provide more features than the analog systems and a higher capacity. Among the digital systems the features are similar; however, as discussed in Section 14.4, CDMA-based systems appear to have a higher capacity than TDMA-based systems. Typically, CDMA base stations are more expensive, but fewer are required because of the increased capacity. Calculating which system is least expensive then becomes highly problematic.

In order to really understand the differences between the systems it is necessary to delve into their technical parameters. These are given in Table 3.2.

When comparing the systems it is important to understand that it is not always the technically best systems that become the most popular. The classic example of this is the battle of video recorder formats between VHS and Betamax. Although Betamax was widely considered to be technically superior, VHS is now the worldwide standard because it was first

Table 3.2
Technical Parameters of the Different Systems

Name	Multiple Access Method	Voice Bandwidth	Carrier Bandwidth	Modulation Type
NMT	FDMA	25 kHz	25 kHz	analog FM
AMPS	FDMA	30 kHz	30 kHz	analog FM
TACS	FDMA	25 kHz	25 kHz	analog FM
GSM900	TDMA	25 kHz (full rate), 12.5 kHz (half rate)	200 kHz	GMSK
D-AMPS	TDMA	10 kHz (full rate) 5 kHz (half rate)	30 kHz	DQPSK
cdmaOne	CDMA	13 kHz or 8 kHz	1.23 MHz	QPSK (downlink), OQPSK (uplink)
PDC	TDMA	8.33 kHz	25 kHz	DQPSK

into the market and particularly because it became the format for which the most software was available. There is a similar argument for mobile radio. Once one system has become accepted in numerous countries, it makes sense for other countries to adopt the same standard because this allows their subscribers to roam, making the service more attractive. Also, the increased sales for this standard provide economies of scale that make the equipment less expensive than other systems. GSM has started down this virtuous route, which often means that even if it has less capacity than cdmaOne, it is still preferred for all these other reasons. This point is discussed in more detail in Chapter 14. Reference works can be found on almost all of these systems and include [1–7].

The GSM system will be described in substantially more detail than any of the other cellular or PMR systems because most of the other systems have much in common with the design of GSM.

3.2   GSM

3.2.1   System architecture

A schematic overview of the GSM system is shown in Figure 3.1. The system is composed of three main elements: the switching subsystem, the base station subsystem, and the mobile. The switching part makes the connection between two users, the base station part controls the communications across the radio interface, and the mobile acts as the transmitter receiver for the user.

The *base station subsystem* (BSS) is shown in Figure 3.2. The BSS consists of a collection of transmitters, known as *base transceiver stations* (BTSs) and the *base station controllers* (BSCs). The BSCs are concentrating points to which a number of transmitters are connected before being connected back along a single line to the switch. This hierarchy simply serves the function of reducing the number of connections into the switch, thus simplifying the role of the switch.

The switching subsystem is shown in Figure 3.3 and consists of the switch and a number of additional databases. In order to find where the mobile is, there are two databases keeping track of the mobile's location: the HLR and the *visitors' location register* (VLR). The basic concept is that every mobile should have an entry in a database alongside which is stored details of its last known location.

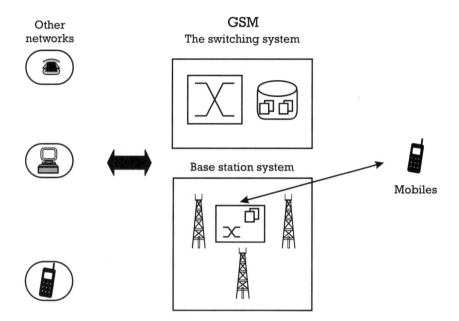

Figure 3.1 Overview of the complete GSM system.

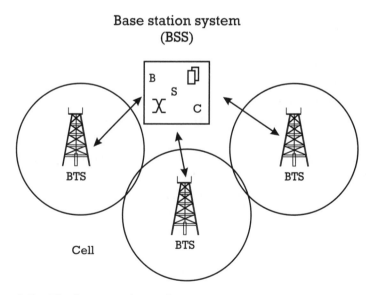

Figure 3.2 The base station subsystem.

In the GSM system, each MSC has its own VLR. When a mobile moves into a different MSC area, the new MSC loads the information

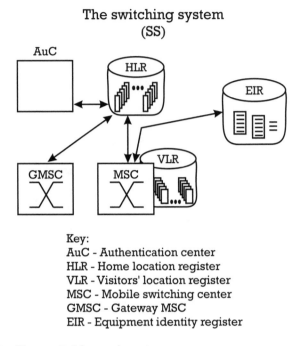

The switching system (SS)

Key:
AuC - Authentication center
HLR - Home location register
VLR - Visitors' location register
MSC - Mobile switching center
GMSC - Gateway MSC
EIR - Equipment identity register

Figure 3.3 The switching subsystem.

from the mobile's HLR into its VLR. This reduces the processing time for calls being set up by storing all the necessary parameters locally and prevents the more frequent interrogation into different networks that would occur if location update messages always had to be sent back to the HLR.

Fraud and theft are key concerns in mobile networks; hence, at registration of the mobile the switch verifies that it has not been stolen. It does so by sending the mobile identity to the *equipment identity register* (EIR), which contains details of all stolen mobiles. The *authentication center* (AuC) authenticates the mobile using the procedure described in Section 2.4.3.

All calls into and out of the mobile radio network pass through the *gateway mobile switching center* (GMSC). The GMSC acts as the point of interface between the GSM network and the PSTN. The PSTN knows that all calls for mobiles in that network should be routed to the GMSC, and similarly, the mobile network knows that all calls destined outside the network should be sent to the GMSC. In the case of an incoming call, the GMSC interrogates the HLR to determine the location of the subscriber and then routes the call on to the correct MSC (which may be in a different network if the mobile has roamed).

3.2.2 Locating a subscriber and starting calls

A basic problem for a cellular system is finding out where the subscriber is located and forming a channel with that subscriber. The various stages of this process are described in the following subsections.

Locating the subscriber   To locate the subscriber, the HLR keeps a record of the MSC area in which the subscriber last registered. The VLR, of which there is one associated with each MSC, keeps a record of the location area in which the subscriber last registered. A location area is a group of neighboring cells that has been "placed" in the same location area by the operator. Each cell in a particular location area broadcasts the same location area code so that the mobile is able to determine when it has moved into a new location area. If location areas were not used and the mobile had to register whenever it moved into a new cell, then the registration traffic would be very high, resulting in a waste of radio spectrum. The mobile is also instructed to periodically send a registration message, even if it has not moved to a new location area, to overcome the cases that the mobile moves out of coverage and so becomes uncontactable or that the location databases crash and need to be rebuilt.

Paging and random access   An incoming call for a mobile is routed to the GMSC of the mobile's home network using the first digits of the mobile number to determine the home network. The GMSC interrogates the HLR that provides the MSC area in which the mobile last registered. This allows the call to be switched to the current MSC (which may be in a different network). The MSC then arranges for paging messages to be sent to all cells in the location area where the mobile last registered. The mobiles listen on the paging channels for a message for their number. If they receive such a message, they instigate a random access procedure that is essentially the same procedure as used for random access in the case that the mobile user presses the send button.

The random access procedure is as follows. The mobile transmits a short message on the random access channel containing only a random number and a brief guide as to the cause of the random access. The message is kept as short as possible to reduce the risk of collision with a random access message sent by another user. The network responds with a message repeating the random number and providing a dedicated channel for subsequent communications. The mobile listens for this random number and, if it hears the number it transmitted, moves to the dedicated channel where it can relay details of the service it requires and the number the user wants to call. If the mobile does not hear its random number repeated, it assumes that the message collided with the random access message from another user and, after a random time (so that the two users who collided are unlikely to retransmit at the same time), it retransmits its request. Once on the dedicated channel, the network sends back to the mobile its own telephone number to overcome the unlikely case that two mobiles were sent a random access message at the same time and with the same random number. If a mobile hears the wrong telephone number, it must leave this channel and retry to send a random access message again. Once on a dedicated channel, if the random access was in response to paging, then the mobile indicates this as the reason for random access and the incoming call can be connected to the mobile.

Handover   Once in a call it must be possible to hand over the mobile to neighboring cells. To provide the necessary input information to achieve this, the mobile is instructed to continually provide measurements of the signal strength from surrounding cells. It makes these measurements during the TDMA bursts when it is not transmitting or receiving. The MSC collects these measurements along with its knowledge of the quality of the existing link. In the case that it decides that a handover is necessary, it tells the new cell to prepare a channel, takes note of the channel it has prepared, sends a message to the mobile telling it the channel to which to move, checks that the mobile appears on this channel, and then deactivates the old channel. The MSC may also decide to hand over mobiles due to traffic congestion. If there is one highly congested cell (say because of a football match just finishing), then mobiles in the cell who could also get coverage from other neighboring cells might be instructed to hand over to reduce the load on the congested cell.

Call routing   The routing of a call in a cellular system was discussed briefly previously. An incoming call (say from a fixed network) is routed to the GMSC in the home network. The GMSC interrogates the HLR to determine to which MSC to route the call. The problem with this approach occurs in the case that a subscriber has roamed to a different country—say, a French subscriber roaming to Germany. A call from a fixed phone in Germany would be routed to the GMSC in France, where it would be routed back to the relevant MSC in Germany. Two international calls have been generated when a national call would have been sufficient. New developments in the GSM standard are underway to overcome this "trombone" problem.

3.2.3 Transmission within GSM

Having considered briefly the GSM system, this section looks in more detail at the transmission of information from the network to the user. Hence, it is concerned only with the part of the system between the BTS and the mobile. For a more detailed description of the GSM system, refer to [1–4]. GSM adopts a TDMA transmission format as discussed in Section 2.4.8. The remainder of this section examines in more detail the manner in which this is divided among the different users.

3.2.3.1 Framing format for speech

GSM uses a TDMA framing format. This means that each radio channel is split into slots and each user is given one of eight slots in which to transmit. The slots are numbered 0, 1, 2, 3, 4, 5, 6, 7, 0, 1, 2, 3, 4, ... The radio channel is normally referred to as a *carrier*. A particular allocation of one in every eight slots is referred to as a *traffic channel*. So each carrier contains eight traffic channels, normally numbered from 0 to 7. Each user who makes a call needs his or her own traffic channel. Each transmission by a user is known as a *burst*. For the user given all the slots numbered 1, the first burst would be in the first slot, the second burst in the ninth slot, and so on.

There is a requirement on the mobile to send information concerning the received signal strength from surrounding cells back to the network in order for the network to be able to decide when the mobile should be handed off to surrounding cells. Transmission of sufficient signaling information to make a sufficiently rapid handoff decision requires a data rate of around 400 bits per second after error correction. This is approximately 1/24 of the data rate provided within GSM. Hence, after every 24 bursts of speech information, one burst of signaling information is sent. In GSM the 26th frame is left empty, for reasons associated with the half-rate speech coding system and channel measurements that will be explained later. The framing structure for the full-rate speech channel is shown in Figure 3.4, where T has been used for traffic, which could be speech; S for signaling; and the 26th burst is kept free. As can be seen, the basic structure is 12 traffic bursts, followed by a signaling burst, another 12 traffic bursts, and a free burst. The transmission then repeats on the 26 burst cycle.

If the half-rate speech coder is employed, then only half as many bursts are required for speech. However, the same capacity is required to transmit measurement information because handoff decisions still need to be taken with the same speed. Users share a slot, as shown in Figure 3.5. Burst 26 is now used to transmit signaling information from the second mobile.

The transmission of signaling information as shown in Figures 3.4 and 3.5 is said to take place on the *slow associated control channel* (SACCH). It is called associated because it is linked with the traffic taking place on that burst and slow because the data rate is relatively slow.

3.2.3.2 Framing format for control purposes

Similarly to the traffic information, it is necessary to send a range of control information to the mobile. Most of this information is of relevance when the mobile is in idle mode, although some is read by the mobile when making measurements of surrounding cells. All the control information is multiplexed together onto a number of control channels, typically one in cells with less than four carriers and two or more in cells with more carriers than this.

The key functions of the control channel are:

▶ To help the mobile locate the control channels;

▶ To provide information as to when the speech and control channels' repetition cycle starts;

▶ To provide information on parameters in the cell;

▶ To provide information on surrounding cells;

▶ To provide paging information;

▶ To allow random access attempts from the mobile.

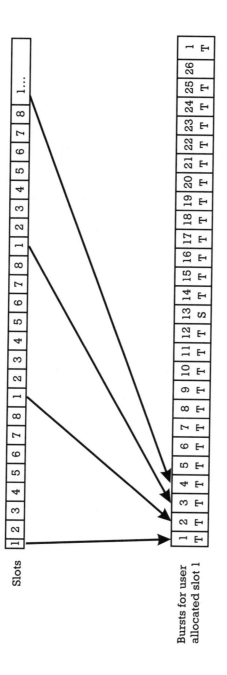

Figure 3.4 The traffic channel structure for full-rate coding.

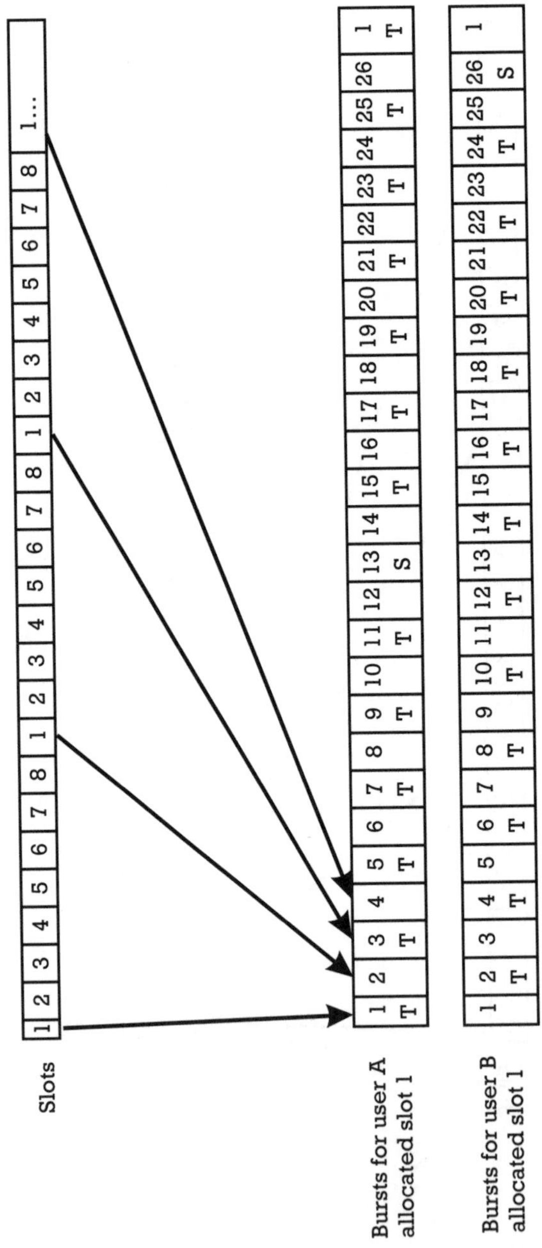

Figure 3.5 The traffic channel structure for half-rate coding.

Helping the mobile find the control channels When a mobile is turned on in a new location, it does not know where the control channels are located. First the carriers being used in the vicinity are detected by measuring the signal strength on each carrier frequency and identifying the strongest carriers. To help the mobile identify the control channels, the network periodically sends a pure sine wave on these frequencies. The mobile listens to each carrier for a period equal to the repetition rate of this sine wave burst and, if it is detected, selects the carrier as the control carrier.

The GSM specifications state that in the next burst on the same channel after the sine wave, synchronization information containing the current position in the framing structure must be sent. The mobile can therefore listen to the next burst and determine at what point the control channel is within its current cycle. The specifications also state that the third burst of the control channel must contain the *broadcast control channel* (BCCH) information, which provides a full guide to the data contained within the complete control channel. Hence, once the mobile has listened to a synchronization burst, it will await the third cycle of the control channel and then decode the position of the complete contents of the channel.

The repetition period of the control channel is 51 cycles. The number 51 has been selected because it has no common multiples with the 26 burst cycle used for the speech frames. Indeed $51 = (2 \times 26) - 1$, a deliberate selection that results in the control channel repeating exactly one burst less than twice the repetition of the speech channels. This enables the mobile to make measurements of the signal strength in neighboring cells, as will be explained later.

Superframes and more The framing structure has a number of levels as shown in Figure 3.6. At the bottom of the figure is the 8-TDMA frame and then the 26 repetition and 51 repetition cycles of the voice and control channels, respectively, as already explained. A single superframe occurs when the speech and control channel restart at the same time. Because 26 and 51 have no common multiples, this means that the superframe occurs every $26 \times 51 = 1,326$ bursts. This is a useful point for synchronization purposes because it helps a mobile switching from a control channel to a speech channel to understand better when to transmit. A hyperframe is comprised of 2,048 superframes and only exists because the hyperframe number is useful as part of the ciphering process.

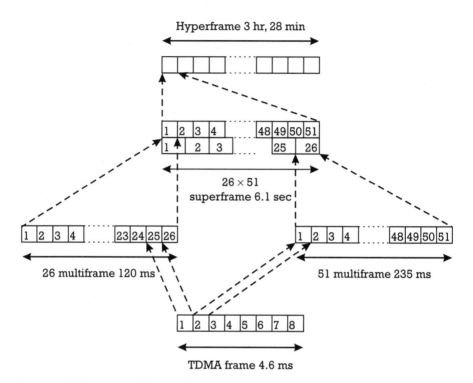

Figure 3.6   Frame hierarchy.

Contents of a burst   The structure of a typical burst is shown in Figure 3.7. At either end of the burst is the provision for the mobile to ramp up and ramp down its power levels in order to avoid an abrupt change in transmitted power that would generate interference over a wide frequency range. Also included is the sounding sequence, as explained in Section 2.4.8, which is used to determine the ISI present in the channel in order to train the equalizer correctly.

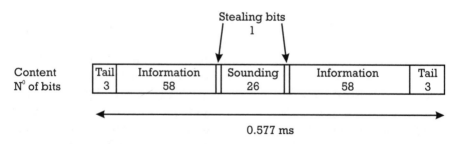

Figure 3.7   Burst structure.

The sounding sequence is placed in the middle of the band such that the time difference between receipt of the sounding sequence and of any of the data bits in the burst is reduced to a minimum. As a result, the changes in the radio channel as a result of the movement of the mobile between the point of receiving the sounding sequence and receiving the data will be minimized. However, the implication is that the mobile needs to store the first half of the burst until the sounding sequence is received and then the burst can be decoded.

The stealing bits indicate whether the information is provided for signaling or user data purposes. In most cases, this information is obvious as signaling information is provided on the signaling channels and user data on the traffic channels. However, in some cases, such as when a rapid handover is required, it is necessary to send signaling relating to the user more rapidly than is possible on the SACCH and a user data frame is "stolen" to make way for signaling. This means that the user data is not transmitted, resulting in a slight break in voice. When signaling information is sent in this manner it is said to be sent on the *fast associated control channel* (FACCH).

Making measurements of adjacent cells   As already mentioned, the mobiles need to make measurements of the received signal strength from surrounding cells in order to provide sufficient information for the network to make handoff decisions. These measurements are performed during the TDMA slots where the mobile is not doing anything else. The mobile receives downlink information from the base station during one slot (say slot 0) and transmits back on the uplink three slots later (slot 3). It also requires a slot to tune its synthesizer to a different frequency. As a result, slots 1, 2, 4, and 7 would need to be used for retuning were the mobile to attempt to measure a different frequency. This leaves the mobile slots 5 and 6 to measure surrounding cells. Since it is unlikely that the slot structure in the surrounding cell is precisely aligned with that in the current cell (propagation delay alone will ensure that alignment is unlikely), the best that the mobile can expect during this time is to read one whole burst from the surrounding cell.

The specifications state that the mobile must read the control channel of the surrounding cells to confirm that it is measuring the cell that the network thinks that it should be. (It is possible that, due to frequency reuse, in some parts of the cell the mobile may receive a signal from a different neighboring cell than that expected by the network.) To do this, the mobile needs to extract the *base station identity code* (BSIC) from the control channels. There is little chance of doing this in the free slots discussed previously because, unless the control channel happens to be on slot 5 in the neighboring cell, the mobile's measurement period will not coincide with the control channel—hence, the reason for a free burst during the 26th burst of the traffic channel. This allows the mobile to listen to all eight slots of one cycle of the TDMA signal, during which time it will be able to receive one burst of the control channel.

The mobile first needs to decode the frequency correction burst and then the synchronization burst in the neighboring cell in order to determine at what point the BSIC will be transmitted. There is only a 1 in 10 chance that the mobile will find a frequency correction burst the first time it listens to the neighboring control channel. Hence, it needs to keep listening on every 26th burst in order to find the frequency correction channel. This is the why control channels have a different repetition cycle from traffic channels—whenever the 26th burst on the traffic channel repeats there will be a different burst on the control channel to which to listen.

With the control channels repeating every 51 cycles, a mobile making measurements in frame 26 would actually read frames 26, 1, 27, 2, 28, 3, ... of the control channels. Hence, an FCCH will be found within 10 repetitions of the 26 slot frame and an SCH will be found 2 repetitions later. Finding the BCCH could take up to 51 repetitions of the 26 multiframe, although on average it will take half of this, or around 3 sec. With an average of six neighboring cells, reading all the BCCH channels will take an average of 18 sec and a maximum of 36 sec. Clearly, if the mobile is moving rapidly through small cells, then this may be problematic.

GSM numbering scheme   Each mobile has at least three, sometimes more, dedicated phone numbers. First, the mobile has the phone number dialed in order to reach it; this is known as the *mobile station integrated services digital network number (*MSISDN). However, if this were the only number and this number were to change, then it would be necessary to return all the SIM cards for reprogramming. The SIM card is a smart card containing details of the subscriber identity, some memory to enable subscribers to store phone numbers, and the encryption algorithm. Having such an easily removable identity element has allowed GSM to benefit from "SIM card roaming" whereby users traveling from Europe, where GSM operates on 900 MHz and 1,800 MHz, to the United States, where GSM operates on 1,900 MHz, for example, can hire a mobile in the United States but insert their own SIM card and hence still make and receive calls using their European number and it is not necessary to recall SIM cards.

In addition to the MSISDN, the user is given another number that is not visible to them and is known as the *international mobile subscriber identity* (IMSI). During an incoming call, the HLR performs the translation between MSISDN and IMSI. Now if a phone number is changed, this database can be updated.

The IMSI is stored on the SIM card. Because the IMSI can be taken out of a phone so easily, it is no good for tracing stolen mobiles. The phone itself also needs a serial number. This number cannot be removed from the mobile and is known as the *international mobile equipment identity* (IMEI).

Spectrum efficiency   The spectrum efficiency of GSM can be calculated as follows. GSM uses TDMA. Each of these TDMA carriers is 200-kHz wide and accommodates eight ongoing calls. So in each 2×1 MHz of radio spectrum there can be $1,000/200 = 5$ carriers, giving $5 \times 8 = 40$ ongoing calls. At least one channel is needed for control purposes, so only 39 or perhaps even 38 ongoing calls can be handled, but the control channel capacity is normally neglected in calculations because it is not greatly significant. In a typical GSM network the cluster size is 12. So in each cell, in each 2×1 MHz, $40/12 = 3.3$ simultaneous calls can be accommodated. This is the spectrum efficiency of GSM, 3.3 calls per 2×1 MHz per cell. In the case of the half-rate coder, this rises to 6.6.

With frequency hopping it is possible to reduce the cluster size to as little as 6. However, the carrier with the control channel on it cannot be frequency hopped because it needs to be transmitted on a fixed frequency to allow mobiles in neighboring cells to make measurements of signal strength. Taking the higher cluster size for these carriers into account, the average cluster size might be between 7 to 9, increasing the capacity by around 30%. Capacity enhancement is discussed in more detail in Section 9.4.

3.3 cdmaOne

cdmaOne is also known as IS-95 and a detailed description can be found in [6]. Here, an overview of the system is provided. cdmaOne is a cellular system based around a similar network architecture to GSM. The key difference arises in the use of CDMA as the air interface technique rather than the TDMA used in GSM. As a result, each user transmits on the same frequency but is given a particular code to use by which their transmissions can be distinguished from those of other users. The same frequency is also used in neighboring cells since signals from each user can be discriminated from interference through the use of their code. This results in a cluster size of 1, maximizing the reuse of the frequencies and giving cdmaOne a higher capacity than GSM.

The downlink transmission is shown in Figure 3.8 and consists of a number of control channels and traffic channels. The control channels include the pilot channel, synchronization channel, and paging channels. These correspond loosely to the *frequency control channel* (FCCH), *synchronization channel* (SCH), and *paging channels* (PCHs) in the GSM system. Speech is encoded using a variable rate voice coder. This generates data between 1.2 kbps and either 9.6 kbps or 13 kbps, depending on the coder used. The data rate depends on whether the user is speaking. Essentially, this is a form of discontinuous transmission where the bit rate falls when the user has nothing to say. However, by varying the rate rather than just having an "on or off" decision, the clipping of speech is reduced and the overall quality is higher. The variable rate results in less interference to other users when this user is not speaking. After the speech coding, error correction coding is applied to generate a data rate of 19.2 kbps. The same rate is generated regardless of the voice rate by duplicating data. This is then interleaved before being scrambled for security and inserting power control bits into the datastream to allow the mobile to regulate its power. The signal is next spread by the allocated 64-bit Walsh codes, resulting in a data rate of 19.2 kbps $\times$ 64 = 1.228 Mbps. Where the voice rate is low, erasure signals are generated so that parts of the signal are not transmitted, reducing interference. The use of 64 Walsh codes allows a maximum of 64 channels; but in practice, if all these channels were used, interference from other cells would be so great that it would not be possible to correctly decode the signals. Finally, the signal is multiplied by the I and Q channel pilot sequences. These differ for different base stations and remove most of the interference that would otherwise occur between neighboring base stations since all use the same Walsh code family.

cdmaOne is unique amongst current cellular radio standards in that transmission on the uplink is different from transmission on

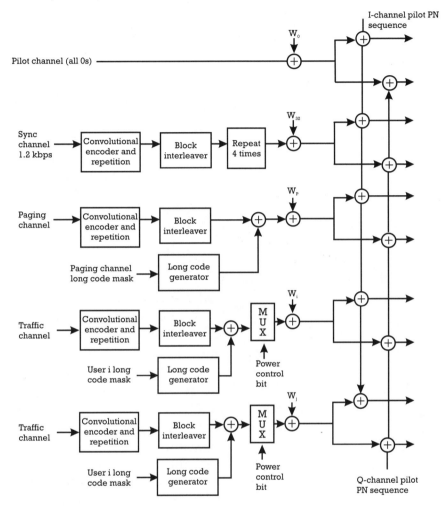

Figure 3.8 cdmaOne downlink transmission.

the downlink. The uplink transmission is shown in Figure 3.9. The same variable rate speech coder is used along with more powerful error correction coding, giving a bit stream of 28.8 kbps (as opposed to the 19.2 kbps used on the downlink). This signal is then spread using the Walsh codes. However, instead of assigning a code to each user, each six bits of data point to one of 64 Walsh codes and this leads to a data rate of 28.8 kbps $\times$ 64/6 = 307 kbps. Hence, more than one user in the same cell could be using the same Walsh code on the uplink at the same time. To

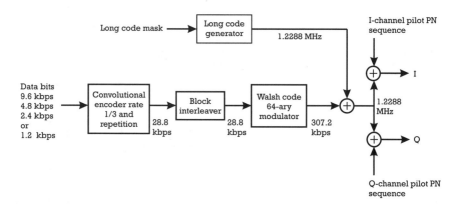

Figure 3.9 cdmaOne uplink transmission.

generate the necessary orthogonality (or near-orthogonality in this case) each mobile generates a unique code to complete the spreading. This code is taken from the pseudorandom code set. Each user generates the same pseudorandom code, termed a "long code" because of its long repetition time. However, each user uses a different offset within this code, resulting in the required near-orthogonality. This multiplication results in a data rate of 307 kbps × 4 = 1.228 Mbps, the final transmitted rate. This is finally multiplied by the pilot I and Q channel PN sequences for that particular base station, again to reduce interference between different base stations.

The reason for the differences between the uplink and the downlink transmissions is due to the fact that the downlink transmission is synchronous whereas the uplink transmission is not. The downlink transmission will be synchronous because all the channels follow the same path from the base station to the user and hence experience the same delays. However, on the uplink, the signals arrive from mobiles that are located at different distances from the base station and hence have different delays associated with them. Truly orthogonal codes, such as Walsh codes, provide the greatest capacity but only under conditions of synchronization. If they are not synchronized, they rapidly lose their orthogonality and the capacity of the system collapses. Nearly orthogonal codes such as PN codes do not provide such high levels of capacity as orthogonal codes, but their capacity does not change much when not synchronized—hence, the greater reliance on Walsh codes for all the spreading on the downlink but only part of the spreading on the uplink. The loss of capacity as a result of the use of lesser codes means that typically the capacity of cdmaOne is limited by the uplink capacity.

cdmaOne capacity is best determined by examination of practical systems rather than analysis. In practice, systems have been able to obtain around 15 mobiles per carrier for mobile systems and up to 30 "mobiles" per carrier for fixed (WLL) systems. With 15 calls per 1.2288 MHz, the capacity per cell per megahertz is given by approximately 12 voice calls per cell per megahertz, which is substantially better than the capacity that GSM is able to provide.

Soft handoff   In GSM, when a handoff decision is made, the mobile disconnects from one cell and then reconnects with another cell. Momentarily it is out of contact with the network and during this time there is a danger that a call may be dropped. cdmaOne overcomes this problem by allowing a mobile to be in contact with more than one base station at the same time. This is possible because the CDMA transmissions in each cell are on the same frequency, and hence mobiles are able to receive transmissions from a number of cells at the same time using a single RF front-end. Normally, the transmissions from other cells are not destined for the mobile and, hence, form interference that is removed during the CDMA decoding process. However, if the network determines that the mobile is approaching a boundary between two cells, it can instruct the new cell to send the same signal to the mobile as the old cell. The transmissions use different codes, so the mobile requires two (or more) baseband receiver units to decode the simultaneous transmission. It can then discard whichever transmission is weakest.

As it moves between the two base stations it will gradually change from discarding most of the frames from the new base station to discarding most of those from the old base station. Eventually, the network decides to stop transmission from the old base station to the mobile. The soft handoff process tends to result in better signal quality during the handoff process with a reduced chance of call dropping.

The disadvantage of such an approach is that two (or more) channels are used by a single mobile during handoff, reducing the overall system capacity by as much as 40% depending on the degree of overlap between cells (and hence the number of mobiles in soft handoff at any particular time). Calculations have shown that the increase in voice quality outweighs the loss in capacity, and soft handover is a mandatory part of the cdmaOne system.

3.4 Other systems

There are two other major digital cellular systems in the world today. These are the IS-136 standard in the United States, often known simply as TDMA, and the *personal digital cellular* (PDC) standard in Japan. IS-136 was designed to increase the capacity of the analog AMPS system by allowing an increased number of voice channels in the 30-kHz bandwidth of a single AMPS channel. To do this it uses TDMA in 30-kHz channels, providing three timeslots. Using quadrature modulation this provides a sufficient bit rate per user. IS-136 features a much simpler logical channel structure than GSM, has field structures within the timeslots that fulfill four logical tasks, and removes most of GSM's complex framing structures. The modulation adopted is $\pi/4$DQPSK. The PDC system is similar to the IS-136 system but lacks the complex interworking functions with a legacy analog protocol.

References

[1]   Mouly, M., and M.-B. Pautet, *The GSM System for Global Mobile Communications*, Paris, published by the author, 1992.

[2]   Tisal, J., *GSM Cellular Radio*, New York: John Wiley, 1997.

[3]   Redl, S., M. Weber, and M. Oliphant, *Introduction to GSM*, Norwood, MA: Artech House, 1995.

[4]   Redl, S., M. Weber, and M. Oliphant, *GSM and Personal Communications Handbook*, Norwood, MA: Artech House, 1998.

[5]   Harte, L., *IS-136 and IS-54 TDMA Technology, Economics and Services*, Norwood, MA: Artech House, 1998.

[6]   Glisic, S., and B. Vucetic, *Spread Spectrum CDMA Systems for Wireless Communications*, Norwood, MA: Artech House, 1997.

[7]   Padgett, J., C. Gunther, and T. Hattori, "Overview of Wireless Personal Communications," *IEEE Communications Magazine*, Jan. 1995, pp. 28–41.

CHAPTER

4

Contents

Private mobile radio systems

If I'd known I was going to live so long, I'd have taken better care of myself.

Leon Eldred

4.1   Introduction

There is a wide range of users who own their own radio systems. These systems are typically known as PMR, although in the United States they are also known under the name *SMR* and by some in the United Kingdom under the acronym *PBR*. Strictly, the term *SMR* applies only to the shared usage of a system; that definition will be used throughout this chapter. The complete wireless professional should know something about this market and these technologies because:

▶ Some mobile radio engineers will end up working in this area;

❯ As will be seen, many PMR users are migrating to cellular, influencing the design and usage of cellular radio systems;

❯ Some of the design principles employed in PMR technologies and networks are instructive for the designers of other networks.

In the early 1980s, a new type of shared PMR system was introduced called SMR in the United States and *public access mobile radio* (PAMR) within Europe. With an SMR system, the network is owned and operated by an SMR operator such as Nextel, in just the same manner as a cellular network is owned and operated by a cellular operator. However, the market base for an SMR operator is typically PMR users who wish to transfer to a shared system. Of course, these users could transfer to a cellular system (and many do); the difference between SMR and cellular is that SMR networks are intended to be more aligned with the needs of PMR users. They typically allow group calls, have no interconnect to the PSTN, and charge users a fixed monthly fee regardless of usage. For heavy PMR users, these systems are often more cost effective than cellular. Life for SMR operators in most countries is getting increasingly difficult as cellular operators target their market and reduce their prices. However, in certain situations, particularly in the United States, SMR operators are experiencing growth, often as a result of their roaming capabilities rather than their group and broadcast calls.

The marketplace   The PMR marketplace is varied, comprising a wide range of users. The total PMR and PAMR market in Europe in 1996 (the last year for which figures were available but these numbers only change slowly) comprised some 4.5M mobile radio terminals. By comparison, the U.S. marketplace over the last few years is shown in Figure 4.1.

The figure shows a market that has been growing steadily at around 8% per year, with SMR taking an increasing share of the market. In 1996, the growth in SMR was around 10% while in PMR it was only around 2.2%. This growth in SMR in the United States is predicted to continue, as shown in Figure 4.2.

The predictions about the U.S. SMR marketplace show an amazing annual growth of 32% with all the growth coming from digital SMR services. This is completely unlike any other country with SMR (or PAMR) operations where subscriber numbers are, at best, rather static.

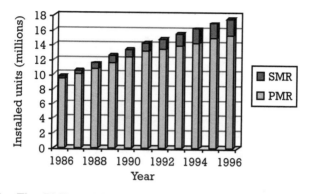

Figure 4.1 The PMR and SMR market in the United States (*Source:* The Strategis Group, London).

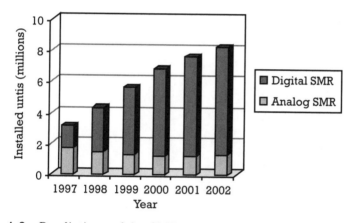

Figure 4.2 Predictions of the SMR marketplace in the United States (*Source:* The Strategis Group, London).

The breakdown of the total number of PMR users within the major European countries is shown in Table 4.1. In some countries the data were not available to segment the user population; hence, only the total is provided. Similar information was not available for the United States, but it is likely that the percentage of users in each category will be similar between the United States and Europe.

The data reveal a sizable European market of over 4M terminals. However, compared to the cellular marketplace, the PMR and PAMR market is perhaps only one tenth of the size. In total, the PMR market is

Table 4.1
European PMR and PAMR Market

Country	Transport	Utility	Industry	Government	Public Safety	Services	Other	Total
Austria	34,000	10,000	26,000	21,000	45,000	—	36,000	173,000
France	113,700	52,000	110,000	110,000	135,000	65,000	66,000	675,000
Germany	—	—	—	—	—	—	—	1,230,000
Italy	—	—	—	—	—	—	—	400,000
Spain	33,000	14,000	47,000	47,000	12,000	51,000	7,800	211,000
Sweden	—	—	—	—	—	—	—	215,000
Switzerland	19,000	9,800	45,000	20,000	47,000	300	3,600	144,000
United Kingdom	180,000	41,000	90,000	97,000	95,000	70,000	277,000	850,000
Total*								4,440,000*

*Also includes data from Denmark, Finland, Hungary, Iceland, Lithuania, Norway, Portugal, Romania, and Turkey.

expected to grow by 3% a year.[1] This is around the level of GDP growth of most countries and suggests that PMR growth is, at best, linked with economic growth.

The total PMR market can be segmented into 10 distinct groups. This segmentation allows the needs and economics of each group to be considered separately. Statistics for each group in the United Kingdom are provided in Table 4.2; the breakdown is similar for the other major countries. In this table, the estimated average cost per terminal per annum to the user includes both capital and revenue components associated with both terminals and infrastructure. For users operating significant control rooms they do not include the costs associated with control room equipment.

The general market segment can be subdivided and summarized as in Table 4.3.

The figures show a total market value of approximately $450M. The public safety organizations provide some 20% of this market, with the rest being divided among a number of smaller user communities.

The choice between cellular, SMR, and PMR   Until the 1980s, the only viable form of mobile communications available was through self-provision of infrastructure and mobiles. PMR is the oldest form of land mobile radio and has grown steadily over the last thirty or forty years.

Despite its growth, however, PMR has its problems. These include:

▸ Increasing congestion of the radio spectrum that has tended to result in a deteriorating grade of service on many systems, particularly in major cities;[2]

▸ The need for users to construct and maintain the necessary infrastructure even though mobile communications engineering may have little to do with the core competencies of their organizations;

---

1. Actually, statistics on the PMR market are rarely gathered, and there are few good predictions available. Some recent work [1] suggests that the PMR marketplace may actually be in a state of decline.

2. Many techniques have been adopted to combat this congestion, including moving from voice to data transmission. These will be discussed on more detail throughout this chapter.

Table 4.2
Summary of Market Segments

	Emergency Services	Utilities	Railways	Local Government	National Government	General PMR	Back to Back	CBS
Includes	Police, fire, ambulance	Gas, electricity, water	Rail and rail police (excludes underground railways)	County, metropolitan, regional, and district councils	Defense, police, immigration	Taxis, couriers, delivery, security, industry, etc.	Mainly on-site users including security, etc.	Commercial common base station services
Number of terminals	95,000	41,000	30,000	82,000	15,000	360,000	170,000	50,000
Cost/user/year	$800–$1,100	$1,100–$1,300	$1,100–$1,300	$500–$650	$800–$1,100	$250–$1,100	$120	$300–$500
Annual value	$96M	$50M	$35M	$50M	$16M	$160M	$21M	$20M

Table 4.3
Summary of General Segment Subdivisions

	Number of Terminals	Cost/User/Year	Annual Value
Taxis/private car hire	142,000	$250–$500	$50M
Motorcycle/ pedal cycle dispatch	10,000	$250–$500	$4M
Bus and coach operators	17,000	$250–$500	$6M
Construction companies	5,500	$250–$500	$2M
Large stores	10,000	$250–$500	$4M
Car industry	2,900	$250–$500	$1M
National	60,000	$800–$1,100	$56M
Other large users	13,000	$250–$500	$5M
Other users	100,000	$250–$500	$35M

▶ Being locked into long system replacement cycles (typically 10 years) that inhibit the ability to ensure that mobile communications are reactive to the changing needs of the organization.

Until 10 years ago these limitations were accepted as a fact of life because there was no real alternative. The advent of public cellular services did provide an alternative. However, there has been only limited migration from PMR to cellular, for a number of reasons:

▶ Cellular has been traditionally viewed as much more expensive than PMR.

▶ Cellular does not offer a number of facilities that PMR offers and that are now central to the operational practices of many users. These include group calls, fast call set up, direct mobile to mobile calls, and various dispatcher facilities.

▶ Cellular does not offer the degree of area coverage that a small number of users require (for example, the electricity companies require coverage along all pylon routes).

▶ Many PMR users, particularly the major users, are extremely nervous about losing direct control over the provision of their mobile communications, fearing that if a service provider does not perform as they would wish, there would be little they could do to improve matters.

▶ PMR handsets are available in specialized forms with special indicators and displays and are more rugged than typical cellular handsets.

With the success of cellular in the early to mid-1980s, it seemed likely that there would be a similar market for shared provision of mobile radio communications for users who were currently operating their own PMR system. However, the SMR market has broadly failed to live up to expectations with many operators merging and recording low growth. The reasons for this are varied but essentially relate to the high costs of providing a national network for a relatively small number of users compared to cellular, resulting in cellular services being both better and cheaper for many applications. It is against this background that operators are considering launching new digital SMR networks, an investment plan that seems difficult to justify.[3]

The GSM operators are looking increasingly well placed to enter the PMR market as the GSM standard continues to evolve. In particular, work by the International Railways to provide the PMR services of group calls and fast call setup within the GSM standard has resulted in the GSM standard containing most of the features required by PMR users.

A key issue in the choice between cellular, SMR, and PMR is the cost. As the tables have shown, the cost of PMR for many users is only in the region of $400 per year compared to around $1,000 per year for the average business user on cellular. Nevertheless, a steady migration away from PMR and toward cellular and perhaps SMR systems seems inevitable in the coming years.

4.2   Simple private radio systems

PMR systems were the first operational type of radio system. Although PMR technology has evolved dramatically over the years, some of the PMR systems deployed are comparatively simple. The most basic PMR systems consist of a simple open-channel transmission system. Here, the

---

3. Indeed, at the time of writing, consolidation was taking place throughout Europe with the number of PAMR operators considering digital networks being approximately halved.

base station is typically located in the building owned by the company. The base station and the mobiles are all tuned to the same channel, typically around 200 MHz in Europe or 450 MHz and 820 MHz in the United States and having a bandwidth of 25 kHz or 12.5 kHz. Whenever an analog voice signal is transmitted it is heard by all the mobiles, which are just functioning as receivers. There is no discrimination between transmissions intended for individual mobiles nor is there discrimination between more than one user of the radio channel. When a mobile wants to transmit back it simply sends an analog voice signal on the uplink channel. If more than one mobile transmits at the same time, then it is likely that neither of their messages will be heard. The coverage area is limited to a single cell and the traffic levels are limited to what can be sent down a single channel, allowing for message collisions.

Such systems are fine for many users. For example, a taxi dispatcher would typically find this system satisfactory. However, problems are caused by a lack of radio spectrum. Because there are more users who would like to operate their own PMR systems than there is spectrum to give them, some sort of sharing needs to take place. One simple means of sharing is to give the same channel to more than one user in the same area. If both users only transmit small amounts of voice, then they will rarely collide and the utilization of the channel will be enhanced. However, problems now occur in that both sets of mobiles will hear all transmissions, despite the fact that only some of these transmissions have been made from their base station by their controller. This is both annoying for the user and can lead to some confidentiality problems (in the case, e.g., where two taxi companies share the same piece of radio spectrum). The next level of complexity was a way of overcoming this problem.

There are a number of ways of ensuring more privacy. One is a *continuous tone-controlled signaling system* (CTCSS). In CTCSS, one of up to 32 subaudio tones is transmitted throughout the duration of the call. These tones range from 67 Hz to 250 Hz. A tone decoder in the receiver mutes the receiver unless the correct tone is present. By giving all mobiles from the same company the same tone but all mobiles from a different company a different tone, users will at least not hear conversations from a different company (however, they will still not be able to use the channel when other users are transmitting). The receiver will prevent a user transmitting onto the channel when there is a signal already on the channel with a different CTCSS tone, which the user will be unable to hear.

A variation on CTCSS, more widely used in the United States, is *digitally controlled squelch* (DCS).[4] Here, a 23-bit sequence is transmitted continuously at a subaudible bit rate. Because demodulation of this signal is moderately difficult without synchronization information, the number of sequences is restricted to 104 different codes. The receiver listens for a particular code and only lifts the mute in the receiver if this code is present.

One of these two techniques is generally used on shared channels to differentiate between users. The users can take this sharing a stage further by noting that since only one base station can transmit at any one time, they might as well share the same base station and reduce costs. Such an approach is known as *common base station* (CBS). The CBS is typically owned by a third party and leased to all the different user groups. At the CBS is equipment that ensures only one user can access the channel at any particular point in time. Although seemingly sensible, CBSs have not been particularly successful for a range of commercial reasons and a lack of radio spectrum.

In the case where there are not sufficient CTCSS or DCS tones to distinguish between the different users, which might be the case where it is required to address individual users rather than teams of users, *selective calling* (SELCALL) can be used. In SELCALL, five sequential tones are sent at the start of each transmission. Each tone is drawn from 1 of 10 codes, allowing up to 10,000 users to be identified. The receiver will only lift the mute if the tone sequence detected is the same as the sequence with which it has been programmed. Group calling is also possible using a wildcard character so that 1234G would call mobiles in the range 12340 to 12349. Because SELCALL takes place at the start of a call, a mobile coming into the coverage area part way through the call would not be able to detect who the call was for and so would not be able to join the call.

In principle, such systems can only work in a single cell since there is no mechanism for changing to a different frequency when moving into a neighboring cell. However, some users have found means to work around this using quasi-synchronous transmission. In this arrangement,

---

4. The "squelch" control in a simple radio is a knob that is turned to set the level at which the speaker is turned on. This means that the user does not have to hear hiss when there is no transmission on the channel but that they will hear voice when it is broadcast, assuming that the noise is below the "squelch level" and the voice above it. Controlled squelch allows the mobile to set the "level" on behalf of the user.

the same signal is sent from neighboring cells on the same frequency. If this is not arranged very carefully, then the effect for a user on the boundary between two cells is to receive a two copies of the same signal with a delay between them. This is akin to receiving ISI and can result in the signal being unintelligible (especially in analog systems where it is not possible to deploy equalizers). The only way to make such an arrangement successful is to ensure that the signals are transmitted from the different base stations at exactly the same time and that even the modulating waveforms are in cycle with each other at the different base stations. This accurate timing can be very difficult to achieve when the signals all originate in the dispatcher office, are sent by landlines to the base stations, and are then transmitted. If the landlines are of different lengths (which will almost certainly be the case), then the signals will arrive at different times at the base stations. It is then necessary to employ delay mechanisms and very precise timing sources to ensure that the signals are all transmitted at the same time. Even once this is done, there tend to be areas where the signal cannot be well received and the system often needs fine adjustments. It has been said that quasi-synchronous operation is more of a black art than a science. However, with the advent of more advanced radio systems, the need for quasi-synchronous operation is gradually disappearing.

Such systems can also only work on one radio channel. Where the capacity of the systems needs to be larger than that which can be offered with a single channel, more complex systems must be employed. These are generally known as trunked PMR systems due to the fact that a number of radio channels are grouped, or trunked, together. As discussed in Section 2.3, Erlang theory shows that trunking a number of radio channels together actually provides much more efficient use of the radio spectrum than the case where each user only has access to one channel. As soon as the need to work on more than one channel is introduced, the complexity of the mobile radio system rises very quickly toward the complexity of a cellular system. Indeed, the distinction between cellular and PMR systems starts to become unclear when PMR systems include trunking.

The most widely used trunked PMR system in the world is MPT1327. This unhelpful name comes from the U.K. telecommunications standard number 1327 that governs its behavior. Because the presence of a standard facilitated competition, MPT1327 rapidly overtook proprietary trunking systems such as Motorola's Smartnet and Smart-zone or Ericsson's EDACS as the preferred option for users. The basic design of an MPT1327 standard calls for a number of 12.5-kHz channels in each cell. One of these channels is nominated as the control channel and provides paging messages for users, telling them of the presence of an incoming call and the channel on which this call will be provided. Most of the principles are the same as for GSM but somewhat simplified. MPT1327 is often known as an analog PMR system because voice calls are carried in an analog form on 12.5-kHz dedicated channels. However, the control channel carries digital information, encoded on the analog chan-nel using *fast frequency shift keying* (FFSK)[5] at a data rate of 1.2 kbps. To find a control channel, the mobile searches each of the known channels for the presence of FFSK signaling and, when it is found, decodes synchro-nizing messages on the channel for confirmation that it is a valid control channel.

The control channel is divided into 106.7-ms-wide slots in which 128 bits can be transmitted. This is split into two 64-bit codewords:

> The *control channel system codeword* (CCSC), which contains a 15-bit code that identifies the relevant mobile or group of mobiles and synchronization information to ensure that the mobiles know when the slots start and stop;

> The address codeword that contains the signaling message; if longer than 64 bits, the next CCSC can be replaced by another address message up to a maximum of four address messages.

When the mobiles transmit on the control channel they send a syn-chronization code and an address codeword followed by their message. There are a number of different message types:

> *Aloha messages* control the random access messages sent by the mobiles.

> *Requests* are sent by the mobiles to request a call.

> *Ahoy messages* demand a response from the radio to ascertain its location.

---

5. FSK is part of the frequency modulation family. FFSK refers to the use of rapid transitions between different symbols using a particular optimal pulse-filtering technique.

▶ *Acknowledgments* are sent to ensure handshaking of messages.

▶ *Go to channel messages* send the mobile to the appropriate traffic channel.

▶ *Single address messages* are used in special cases where it is necessary to send the radio serial number.

▶ *Short data messages* allow the transmission of short bursts of data up to 184 bits long between users.

▶ *Other* includes, for example, network information.

More information about the MPT1327 standard can be found in the MPT1327 specification, [2] or [3].

4.3 TETRA

4.3.1 Introduction

TETRA is intended to be one of the new generation of digital PMR equipment that also includes APCO25 as the equivalent U.S. standard. Standardization has taken place during the 1990s within ETSI, and in 1998 the first TETRA product was starting to appear. TETRA is designed to be a multipurpose mobile radio system for a wide range of users including emergency services, utilities, airports, and PAMR operators. Although possible technically, it is thought unlikely that TETRA systems would be deployed by small users such as taxi companies since this option would be expensive. However, such users might become subscribers to a shared TETRA system. The TETRA standard is complex and extremely comprehensive. A fully specified TETRA radio will be more advanced than a GSM handset and able to perform a wider range of functions.

TETRA comes in two basic forms: the *TETRA voice and data system* (TETRA V+D), which is designed for general purpose use and provides voice and data circuit-switched connections, and the *TETRA packet data optimized* (TETRA PDO) standard, which is designed for packet data applications. Both standards operate over the same physical TETRA network but are designed to be optimized for different tasks. In order to describe TETRA in more detail this section will concentrate on the V+D standard.

This also includes a direct mode capability whereby one terminal can talk directly to another without the need for infrastructure.

In order to fully understand the capabilities and design of a TETRA network, it is important to know that much of the standardization work was driven by the police forces throughout Europe. As a result, TETRA has a range of facilities designed primarily for public safety applications. Some have questioned whether this very high level of specification has left TETRA over-specified for commercial applications. This remains to be seen.

An overview of a TETRA network is shown in Figure 4.3. As can be seen, much of the architecture is similar to that of GSM, although there is the additional mobile to mobile mode shown at the bottom of the figure that is not available within GSM. Given that TETRA has many of the functions of a cellular network, this similarity should not be surprising.

4.3.2 System operation

The operation of the TETRA system is similar to the operation of GSM. On power-up, the radio searches for a control channel and registers itself

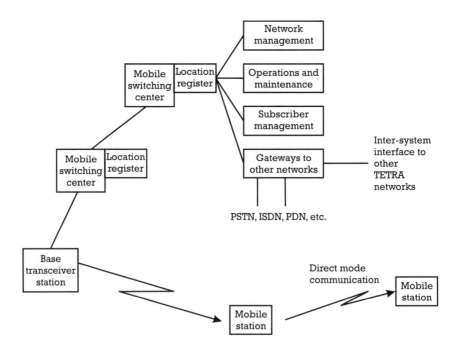

Figure 4.3 Overview of a TETRA network.

onto the network. The coverage area is divided into a number of location areas, the mobile station only being paged in those cells in its current location area. Mobiles need to reregister upon moving into a new location area. Unlike GSM, TETRA does not have a handoff procedure. Instead it makes use of a fast call reestablishment procedure such that when it moves to another cell, the call can be rapidly reestablished in the new cell. The break in communications can be made to be virtually as short as for a handoff in GSM.

TETRA has a very high level of security, commensurate with its deployment in emergency services. It allows both air interface encryption and end-to-end encryption of information. The actual encryption process, including the use of a temporary identity, is very similar to that used in GSM where a random number is sent to the user that they combine with their secret number and return the result; however, the GSM system cannot provide end-to-end encryption.

The random access process at the start of a call is also similar to GSM. The base station sends information as to when random access is allowed that is monitored by the mobile. The base station can allow different access classes of mobile access at different times. There are four access classes, A to D, and the mobiles are assigned one of these four access classes in a manner that can be defined by the operator. The access classes can also be linked with message priorities so that at any point the lower priority messages can be blocked. Once a particular access class is allowed, a mobile picks a random number of slots after the first allowed access and makes an access attempt using a slotted-Aloha protocol. A successful access is acknowledged with an ACCESS-ASSIGN message from the base station that contains details of the reserved channel that the mobile should use for further communications. Unlike GSM, the initial random access contains a substantial amount of information. Although this increases the probability of random access collisions, it means that for a successful message the call setup time can be reduced because less signaling is required at the start of the call. Rapid call setup has been judged to be essential for public safety use. After the ACCESS-ASSIGN message, the mobile is given a short transmission resource to allow it to linearize its power amplifier by transmitting a short burst. It can then move directly to transmitting voice frames.

Once in a call, TETRA offers a range of handoff mechanisms. There are five types of cell reselection depending on:

▶ Whether a call is currently in progress and if so whether it is a group call or individual call and whether the mobile concerned is the transmit mobile or the receive mobile (i.e., listening to a group call);

▶ Whether the mobile had sufficient time to gain information on the new cell;

▶ The grade of call reestablishment supported by the infrastructure.

The categories of call reestablishment are:

▶ Undeclared, where the mobile is not currently in a call and simply makes an idle mode reselection decision;

▶ Unannounced, where the mobile is currently engaged in a circuit mode call but does not have the opportunity to inform the infrastructure that it is changing cell; in this case, it must make a random access in the new cell and detail its situation, and the break in communications could last for more than 1 sec;

▶ Announced type 3, the same as unannounced;[6]

▶ Announced type 2, where a mobile engaged in a circuit mode call informs the present cell of its intention to find service on a new cell and then negotiates directly with the new cell on its *main control channel* (MCCH);

▶ Announced type 1, where a mobile engaged in a circuit mode call informs the current cell of its intention to find service on a new cell and of its preferred cell; negotiations for access to the new cell are performed via the present serving cell, channel allocations on the new cell are issued by the present serving cell, and this process amounts to seamless handover.

In order to provide the greatest flexibility, the TETRA specification offers a number of modes of service. Some of the key modes follow.

Normal mode   The normal mode is intended to cope with the majority of installations where there are typically four or five carriers in use on each site. Here, the first timeslot on the main carrier is set aside to

---

6. It is not clear why there are two seemingly identical categories. This is probably a historic "accident" of standardization.

form the control channel. All mobiles not in a call listen to this control channel. The base station transmits on all downlink slots of the main carrier in order to allow mobiles in neighboring cells to measure signal strength.

Extended mode   In larger installations, more than one control channel may be required. A common secondary control channel with the same functionality as the main control channel is transmitted but is only used by a subset of the users.

Minimum mode   The minimum mode is used in low-traffic density areas where there is typically only a single carrier deployed. In this mode, all the timeslots on the main carrier can be set aside for traffic. Hence, only the control channel frame, frame 18, is available for control channel purposes. Mobiles must be aware of when the base station has entered minimum mode and not attempt to decode traffic information.

Time sharing mode   In the time sharing mode, a carrier is shared between a number of base stations by only being used for some of the time. This is only suitable for very low traffic density areas.

4.3.3   Technical parameters

TETRA operates in a TDMA frequency division duplex mode. Each carrier is 25-kHz wide and provides 4-TDMA frames. The signal is modulated using $\pi/4$ *differential quadrature PSK* (D-QPSK) modulation[7] that allows the encoding of 2 bits/Hz. After allowing for guard bands, this results in an overall transmission rate of 36 kbps per carrier or 9 kbps per user. However, it is possible to assign more than one slot to one particular user, allowing them to increase their data rate to 32 kbps. Transmission and reception slots are staggered by 2 time slots so that mobiles do not need to transmit and receive at the same time. Further, uplink and downlink slots need not be paired but can be assigned to different users. The basic TETRA framing structure is shown in Figure 4.4.

---

7. QPSK transmits 2 bits per symbol using a constellation of four points of equal amplitude. D-QPSK uses the difference between transmitted symbols rather than the absolute value to encode the information. $\pi/4$ D-QPSK changes the constellation by 45 degrees each symbol period, resulting in a continually changing waveform that helps clock recovery at the receiver.

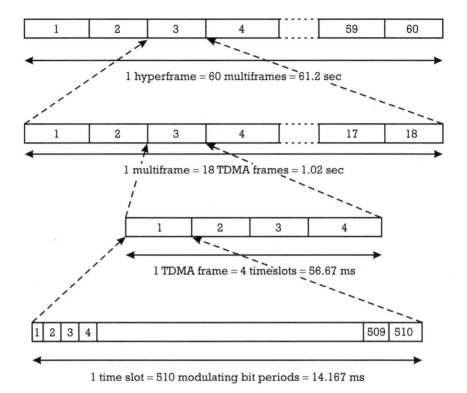

Figure 4.4   The TETRA framing structure.

Unlike GSM, there are no additional control channel structures to complicate the framing structure. The control channel forms the 18th frame of every multiframe, allowing control messages to be sent along with the voice call information. This is called the SACCH and is equivalent to the SACCH in GSM. The normal burst structure involves a short period for ramping up in power, the first half of the user data, the training sequence, and the second half of the user data. After allowing for the error correction information, the user can send up to 408 bits per time slot, allowing a maximum data rate of 7.2 kbps. One frequency pair per site is designated to carry the MCCH. Normally slot 1 on this channel is used for control purposes.

TETRA offers a number of area coverage techniques.

▶ *Single-site wide area coverage:* This is a straightforward single base station on a high site providing good coverage but poor spectrum efficiency as a result of the large cell size.

▶ *Cellular channel reuse:* Normally used in a medium- to high-traffic density networks, a cellular-type deployment is adopted with a clustered arrangement of frequencies as introduced in Section 2.3.

▶ *Quasi-synchronous transmission:* Used in areas that cannot be covered by one site and where a limited number of radio channels are available, special synchronization techniques are necessary for the base station and the technique by which the control center is connected to each of the base stations. A special class of mobile may also be required with a complex equalizer.

▶ *Time sharing transmission:* This is the same as the time sharing mode discussed previously where a carrier is only deployed in a given cell for some of the time and is rotated around a number of cells.

As can be seen, with the introduction of TETRA, the differences between cellular systems and PMR systems have become very small indeed. The implications of this will be discussed further in Section 14.2.

4.4 Other systems

There are two other key digital systems under development. One is the APCO25 system being developed in the United States for public safety purposes, which uses FDMA but otherwise provides very similar functionality to TETRA. The other is Motorola's iDEN, which is a proprietary system used by Nextel to provide PAMR service across the United States. Because of their similarity to TETRA, they are not described further here.

References

[1] MPT1327, MPT1327 Specification. Available from the Radiocommunications Agency; see www.open.gov.uk/radiocom/rahome.

[2] Holbeche, R. J., ed., *Land Mobile Radio Systems*, London: Peter Peregrinus, 1985.

[3] NERA, Smith, *Review and Update of 1995 Economic Impact Study*, available from the U.K. Radiocommunications Agency, 1997; see www.open.gov.uk/radiocom/rahome.

CHAPTER

5

Contents

Other mobile radio systems

I find that a great part of the information I have was acquired by looking up something and finding something else on the way.

Franklin P. Adams

5.1   Introduction

Although cellular and PMR make up the two types of mobile radio system that the complete wireless professional needs to understand, there are a number of other types of mobile radio system of which the communications professional should also be aware. These include cordless systems, WLL systems, video distribution systems such as the *local multipoint distribution system* (LMDS), and satellite systems, and are introduced in this chapter.

The key issues discussed in this chapter are as follow.

▸ Cordless phones are designed for indoor and office applications and do not transfer well to outdoor, cellular-type systems. They are typically much simpler than cellular phones using lower power transmission and higher bit-rate speech coders.

▶ WLL systems are designed for fixed communications in situations where it is easier, cheaper, or more advantageous than wireline connections and are often based on cellular or cordless technologies. There is a wide range of different environments into which WLL can be deployed and hence the need for a range of different technologies. WLL systems need better voice quality and a wider bandwidth than cellular and cordless systems.

▶ Video distribution services provide an alternative TV distribution mechanism to satellite or terrestrial broadcast but typically need to work at high frequency bands above 20 GHz in order to find sufficient spectrum and normally only provide one-way communications.

▶ Satellite systems are the only way to provide global coverage. In order to provide acceptable quality, *low Earth orbits* (LEOs) are required but result in the need for numerous satellites and hence expensive systems. Whether there is a sufficient market for such a system remains to be seen.

5.2 Cordless systems

5.2.1 Overview of cordless telephony

Considerable confusion exists about cordless telephones because they have been used in a wide range of different roles. To understand cordless properly, it is necessary to look at each of the different roles. We discuss three roles here, but its role as a WLL technology in more detail in Section 5.3.

Mobile use within the home   The initial application of cordless phones was as an extension to the fixed line in a home. The cordless base station is plugged into the home phone socket and provides a radio link to the handset. For such a deployment to be successful the phone needs to be inexpensive and able to operate autonomously. The cost is kept down by simple design and by a restriction in range to around 200m while autonomous operation is provided by the base station performing dynamic channel allocation whereby it seeks the radio channel with the lowest interference and uses that for its operation.

Mobile use within the office   An obvious extension to using cordless phones as a way of providing mobility in the home is to do the same thing in the office. Most offices have a *private branch exchange* (PBX) that connects wires from each user's fixed telephone. The cordless system removes these wires, replacing them with a single connection to the cordless office system. This then connects to a number of base stations around the office. To some extent this is like a minicellular system within the office; however, the system is typically much simpler, with automatic frequency allocation, typically little in the way of location updating for a user, and a much shorter range. Nevertheless, handovers and other cellular features are provided and the user gains the benefit of mobility within the office. To take a cordless home phone design and translate it to a cordless office design requires the addition of handover, the ability to control a number of base stations, and the capability to dial internal extensions that can be routed through the PBX. However, the basic parameters of the system remain unchanged.

As a limited mobility cellular system   The terminology here might appear slightly strange since cellular systems have limited mobility, typically not providing coverage across all parts of the country. However, "limited mobility" has come to be used to describe a system offering less mobility than cellular systems. The use of cordless phones to provide a pseudocellular system is often known as telepoint. Telepoint is the deployment of cordless base stations in streets and public areas to provide a low-cost service that, at first, appears similar to cellular. The key difference is that the base station range is so low that coverage can only be provided in high density areas, unlike cellular where more ubiquitous coverage is provided. In early systems there was also the disadvantage that incoming calls could not be received. Given that this reduction in range is a major disadvantage, it should not be surprising that telepoint systems have broadly failed. The United Kingdom launched four networks that failed, as did networks in France, Germany, and a variety of other countries. Typically, the lack of coverage compared to cellular makes the service unattractive. Telepoint has been successful in a very limited number of places, namely Singapore, Hong Kong, and more recently Tokyo. These are all very high density cities where the populace rarely leaves the city areas. In such cities, widespread deployment of cordless base stations can be worthwhile because of the high user density and relatively good coverage that can be provided. The high capacity that results from using very small cells allows many more users onto the network than is possible for a traditional cellular system,[1] allowing call charges to be lower than cellular, making the service an attractive proposition. It is unlikely that there are many other cities in the world where cordless will be successful in a telepoint application and its main sales will remain as home and office phones.

The key cordless technologies of DECT and *personal handiphone system* (PHS, the Japanese cordless standard) are now described in more detail.

5.2.2 Digital enhanced cordless telephone

5.2.2.1 Introduction

The DECT specification was developed by ETSI to provide an air interface for cordless telephony that is able to support a wide range of services. DECT standardization began in 1988 and was still ongoing in 1998, as new modes, such as WLL, were added into the DECT standard. DECT is now able to support residential cordless applications, business cordless applications, radio *local area networks* (LANs), and WLL systems. In order to support this wide range of applications, DECT was designed with the following basic features:

▶ A very high voice call capacity of up to 10,000 Erlangs/km^2 per floor in indoor environments with a speech quality similar to wired networks;[2]

▶ High data rates with several megabits per second overall network capacity;

▶ High-capacity base stations but allowing the use of simple radios;

---

1. Although not with cellular systems that deploy microcells—these can provide greater capacity than cordless systems.

2. Although DECT can accurately claim to be able to support such high-traffic densities, this is not primarily achieved by a highly spectrum-efficient radio system but by the capability to install a large number of cells in proximity. Indeed, on a typical measure of spectrum efficiency such as channels/MHz, DECT performs poorly compared to cellular technologies.

▶ A system design that is capable of avoiding the need for frequency planning;

▶ Support of ISDN services.

The DECT system can be represented by a layered model conforming to the OSI seven-layer model. In such a representation, the lowest layer is the physical layer. Above this sits the medium access control, and above that the data link control. The highest level in the DECT specifications is the network layer, conforming to the OSI layer four. Each of these layers is discussed in the following subsections.

5.2.2.2 Physical layer

DECT divides its spectrum allocation of 1,880 MHz to 1,900 MHz into 10 discrete frequency bands, each 2-MHz wide. Each of these frequencies is then accessed using TDMA. A diagram of the TDMA structure used by DECT is shown in Figure 5.1.

This figure shows that the TDMA frame of 10 ms is divided into 24 slots, of which nominally 12 are used for the base station to mobile direction and 12 for the mobile to base station direction. However, it is possible to reverse the use of some of these slots in a dynamic fashion if

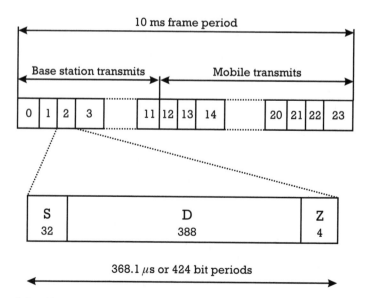

Figure 5.1 The framing structure used within DECT.

the traffic to be carried is asymmetric. Each burst contains three fields. The S-field contains synchronization information in the form of a Gold code that aligns the timing and ensures that mobiles are aware of the current position within the burst structure. The D-field contains the user data, while the Z-field just repeats the last four bits of the D-field. This allows early detection of a collision that may occur where more than one DECT system is trying to share the spectrum in the same area.

Data is modulated using two-level FSK with a BT product of 0.5 and a modulating data rate of 1,152 kbps. This scheme permits the use of a simple receiver with noncoherent detection and without the need for an equalizer.

5.2.2.3 Medium access control

The *medium access control* (MAC) layer has three functions:

1. To allocate and release physical resources according to the requests of the upper layers;

2. To multiplex the signaling and user information channels onto the physical timeslots;

3. To ensure secure transmission through the use of error correction coding.

Allocation of the physical resources   One of the key features of DECT is the use of *dynamic channel allocation* (DCA) whereby any base station can select any of the 10 radio channels to use at a particular instance and with no control from the switch. This is essential in cordless systems where users may own both the base station and the mobile and hence no single body has control of all the base stations using the same frequencies. Channels are selected based on the measured interference levels, with the base station choosing the channel with the lowest observed interference at any point in time. Regardless of whether there are any users, each base station must always transmit on one carrier, known as the beacon frequency. This allows mobiles to scan for the channel from the closest base station and to lock onto this channel. The channel is then used for paging messages. The beacon frequency is also used for any ongoing calls, so it is not necessary to reserve a frequency solely for beacon transmissions. Its function is very similar to the BCCH channel in the GSM system.

Multiplexing of the logical channels   There are a number of different channels in the DECT system, as shown in Table 5.1.
Many of these channels are analogous to those in GSM; for example,
the I channels are equivalent to GSM TCHs, the C channels are equivalent
to the SACCH and the FACCH, the P to the PCH, and the Q to the BCCH.

A standard MAC packet uses the 388 bits available in the TDMA burst
in the manner shown in Table 5.2.

It can be seen from this table that the paging information can be sent
as part of a frame destined for a particular user. Because this paging information can be sent to any mobile, in a divergence from GSM, a burst can
have a dual purpose and so mobiles need to listen to all downlink bursts,
regardless of whether the remainder of the frame is destined for them.
Since each frame repeats every 10 ms, the M, N, P, Q, and C channels have
a data rate of 4.8 kbps, while the user data is transported at 32 kbps.

Table 5.1
The Different Channels in the DECT System

Name of Channel	Description
I	Information: This can be carrier in either a nonprotected or a protected mode.
C	Control channel for signaling, which can be slow signaling sent concurrently with the data or fast signaling sent by stealing data frames
M	MAC internal control channels
N	Identity channels
P	Paging channels
Q	System information broadcast channels

Table 5.2
The Usage of the 388 Bits Within the Burst

Number of Bits	Usage
48	Used for either the M, N, P, Q, or C channels
16	Error correction bits (termed CRC)
320	User information or fast control channel signaling
4	Error control bits (termed X)

Error control   If C, P, or Q transmissions are sent, they are protected with a 16-bit *cyclic redundancy code* (CRC), a form of block coding. However, the coding is only used to detect errors, not to correct them. If P or Q transmissions are detected to be in error they are ignored; however, if C channels are in error they will be requested to be retransmitted. The 4 X bits are used to detect errors in the user information. They are far too few bits to provide any strong error protection but can indicate whether there has been large scale interference with the burst.

5.2.2.4   The data link control and network layer

The *data link control* (DLC) provides higher levels of error correction for the user data and ensures that handover is performed correctly. The *network* (NWK) layer handles the routing of incoming calls by tracking the location of the users in a similar manner to the GSM location registers.

More information about the DECT system can be found in [1, 2].

5.2.3   Personal handiphone system

PHS is a cordless standard developed initially for use in Japan, but which has also found application in a number of other Asia-Pacific countries. The standard was developed by the Japanese standardization body in conjunction with Japanese industry.

Radio system   PHS uses $\pi/4$ shifted QPSK with a roll-off factor of 0.5. This provides similar simplicity as for DECT and equally means that equalizers are not required. Like DECT, the multiple access method is TDMA but using only four timeslots per carrier. TDD is also used. Carriers are spaced at 300 kHz apart and provide a bit rate of 384 kbps. By the time allowance is made for additional information, and for the TDD duplex, resulting in a one-way data rate of only half of the total rate, each user is left with a 32-kbps data rate allowing ADPCM speech coding. Antenna diversity is employed at the base station, and by selecting the best antenna during the base station receive mode and retransmitting on this antenna during the base station transmit mode, diversity gains can be achieved in both directions.

The frequency band assigned in Japan is 1,895 MHz to 1,918 MHz, of which 11 MHz is allowed for both public (i.e., telepoint) and private (i.e., residential cordless) use, while the remaining 12 MHz is set aside for purely public use. In total, 77 channels are available, of which 6 are set aside for control channel purposes. Handover capabilities are provided. A form of dynamic channel allocation is used, although different techniques need to be adopted for control channels compared to traffic channels. For the control channels a form of quasi-static autonomous assignment algorithm is used, whereas for the traffic channels a standard DCA is adopted.

Radio channel types There is a similar set of radio channel types as in DECT and GSM. *Control channels* (CCHs) include the associated control channel, common control channel, and broadcast channel that provide system information. PCHs provide paging information while *specific cell channels* (SCCHs) provide information related to call setup. *User packet channels* (UPCHs) provide packet data transmission, and *traffic channels* (TCHs) carry the user information.

More details about PHS can be found in [1].

5.3 Wireless local loop systems

5.3.1 Introduction to wireless local loop

WLL is the provision of a cellular-like phone without mobility. This may seem a strange concept given that one of the key benefits of a cellular phone is mobility. To understand why a mobile phone would be provided without mobility it is necessary to separate the two cases of developing and developed countries.

In developing countries, very few people have a telephone. In African countries, less than 2% of the population has access to a phone. This low level of penetration is typically because the users cannot afford to have a phone installed in their house. Much of the cost of installation of a standard phone is associated with the cost of installing the copper cable required to link the home to the switch—a process that has taken 100 years in developed countries. Wireless systems offer an opportunity to provide a phone connection more quickly and at lower cost. By restricting the position of the receiver to being mounted near an appropriate window in the house, the range of the cell can be extended since there is less blocking between the phone and the base station than would be the case with a mobile phone. This allows fewer base stations to be used and hence the system to be deployed more cheaply than would be the case with a mobile system. There is also no need for the added complexity of handover.

In developed countries the situation is quite different, with most homes having phones. However, some countries are interested in offering a phone service in competition to the PTO and see wireless as the least expensive means of providing the service. In these countries the user could afford cellular systems; however, there are a number of other problems with cellular when compared to landline:

▶ The voice quality is typically inferior.

▶ The data rates provided are typically lower.

▶ The call costs are typically more expensive.

An operator wishing to deploy a wireless service has to make use of a technology that overcomes these problems. The first is overcome using a higher voice coding rate, and the second using technologies providing higher data rates. Overcoming the third is broadly achieved by only providing coverage in city areas and hence requiring many fewer base stations than would be required for a cellular system.

In principle, then, a WLL system is one employing a cellular-like technology but where the subscriber unit is fixed, providing an equivalent of a wireline telephone, replacing the "local loop" segment between the exchange and the subscriber's home with a wireless link.

In recent years, some of the WLL systems being deployed have provided a combination of both fixed and mobile services and it is worth examining in more detail why this should be the case. In principle, the operator of a cellular network could provide a WLL service as well by simply providing users fixed versions of their mobile phones. This seems somewhat pointless since they could equally be supplied with mobile phones, allowing them to gain the benefit of mobility. However, the user will only accept a replacement home phone if their call costs are as low as a wireline phone whereas cellular calls are typically more expensive. To overcome this, the mobile operator needs to use WLL as a means of providing the user with a lower tariff. Again, this seems pointless since the cost to the mobile radio operator of carrying the call is the same regardless of whether the call comes from a fixed or a mobile phone. What is actually happening here is that the mobile radio operator could carry all the calls from mobile and fixed phones at the wireline rate but to do so would dramatically reduce their profit. By maintaining a high price for mobile calls and a lower price for fixed calls they retain high margins on the mobile calls while increasing the overall traffic volumes on their network.

So in summary, WLL is the use of radio technology to provide a fixed connection to the home. There are many different technologies that can be used, and these are discussed in the remainder of this section. Before looking at the different wireless technologies, the different access technologies including copper, cable, and fiber optic are considered so that the overall access marketplace can be better understood.

5.3.2 Access technologies: radio and cable

The cellular systems discussed in previous chapters have no competition from other modes of information delivery—if communications on the move is required, then cellular systems have to be used. This is not the case with WLL, despite the fact that WLL systems are based on cellular principles. There are numerous other means of making connections from the home to the PSTN and this set of connections is generically called the access network. Although there are a number of different WLL technologies, which compete with each other, many perceive that the key competition to WLL comes from copper cable. Hence, it is important to understand a little about the world of access technologies for fixed applications even though they are clearly not wireless. This section provides a brief overview of the available technologies.

The number of access technologies is continually increasing and includes copper cable, cable TV systems, cellular, cordless, satellite, and even information sent over power cables as has been recently suggested. Driving the increase in the number of access technologies is the concept of convergence. Although meaning different things to different people, broadly convergence predicts that a range of different media, including voice, computer data, and video, will be transmitted along the same path, possibly in an integrated form. What is now happening is that the delivery channels used by each of these three applications are being expanded so that they can become the preferred channel for converged data. This has resulted in competition between channels that previously had little in common—for example, between the copper laid by the PTOs and the coax used for cable television. The complexity of the situation has been exacerbated by liberalization of telecommunications and by tariffs that do not reflect costs—for example, international calls cost little more to deliver than local calls but are charged at a much higher rate.

The main contending access networks are shown in Figure 5.2. This is a figure full of acronyms and buzz words that will be explained in the remainder of this section. The figure itself is confusing and full, which well reflects the confusing situation currently facing providers of access networks. The figure shows approximately as many access networks moving from left to right (i.e., from voice toward computing and television) as are moving in the opposite direction.

Each of these access mechanisms is now described. At the end of each description a brief prediction is made about the role that the mechanism will play in the future of telecommunications.

The copper infrastructure   The copper infrastructure, comprising twisted-pair copper cables installed and owned by the state telephone company (the PTO), has formed the traditional access infrastructure for almost 100 years. Until recently it was ill-suited to

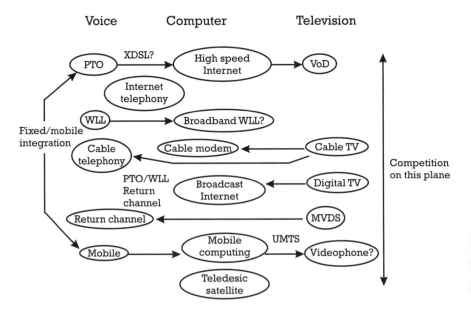

Figure 5.2   Different access technologies and convergence.

anything other than voice due to its low bandwidth (only around 3 kHz). However, a number of recent advances have started to change this. Voice-band modems can now achieve rates of up to 56 kbps, which provide slow but viable computer access. ISDN provides rates of up to 128 kbps, offering the potential of video, albeit of a low quality. The role that this copper will play in the future, though, will be dominated by the success or failure of *digital subscriber line* (DSL) technology, which promises a massive increase in the data rate passed through copper wire.

A whole host of different approaches to DSL systems have been proposed with names such as *asynchronous digital subscriber line* (ADSL), *high-speed digital subscriber line* (HDSL), *very high-speed digital subscriber line* (VDSL), and doubtless more to come. The acronym *xDSL* refers to all these approaches.

ADSL is a complicated means of getting much higher data rates down telephone lines—being a wire-based technology, it falls outside the scope of this book. Such a high-speed technology would allow the national tele-communications company to deliver services such as video on demand through existing connections. ADSL is currently under development, and it is still not clear whether it will be successful.

Internet telephony Internet telephony is currently an application that is concerning many existing operators. The concept is to use advanced speech coders to compress and segment the voice and then send the packets over the Internet. The advent of this technology is mostly artificial due to the fact that all Internet calls are charged at local rates despite the fact that many result in international flows of data. This, in fact, reflects the true cost of telephony provision, where the local loop section represents the vast majority of the cost, and the national and international trunk contribution is almost insignificant. However, fixed operators persist in inflating international prices in order to increase profitability and in some cases subsidize the local loop. If Internet telephony works (and it is unlikely to for long, even if it does work at first, as the network simply becomes swamped by voice as well as the current heavy data load), the net result will simply be that fixed operators will rebalance their tariffs so that Internet telephony is no longer worthwhile. The only advantage then will be the ability to link voice with exploring a site so that a user could contact the operator of the site they are surfing for help if required.

In the longer term, packet-based systems may gradually replace circuit-switched systems, which will improve the efficiency of use of the access networks but not necessarily change the role of any network. Finally, it should be remembered that all Internet telephony calls take place over existing access networks, typically the copper infrastructure. All that changes is who gets paid and how much.

Wireless local loop   WLL provides a similar level of service to wireline systems at similar or lower data rates. However, it brings a number of key advantages.

▶ It is typically less expensive to install wireless than wireline systems.

▶ Installation is more rapid.

▶ Installation can be delayed until the moment when the subscriber requires service.

▶ The economics of WLL remain similar even in low-density areas, whereas the cost per subscriber for wireline systems increases as the distance between the subscribers increases.

As a result, WLL is being considered as an alternative to deploying wireline systems in many countries; and some analysts have predicted that, by the year 2005, there will be more wireless lines than wired lines being installed. More details about WLL can be found in [3].

Cable TV   Cable TV systems use buried coaxial cables to provide a broadband service to their customers. These systems were initially one-way TV distribution systems that provided around 50 TV channels using analog modulation. At the receiver end, all that was required was to change the frequency of the signal on the cable to one that was within the bandwidth of the TV receiver. Subsequently, cable operators decided to increase their revenue by offering telephony. This caused major problems for many of the older networks because one-way amplifiers were in use, which meant that the signal could not be easily carried back from the subscriber. Some of these operators even added a second, telephony network, using separate twisted pairs alongside their coaxial network. Newer cable networks have overcome these problems by deploying two-way amplifiers and carrying voice traffic over a low-frequency carrier within the coaxial cable. Once this form of connectivity is in place, the transmission of computer information is relatively straightforward, making use of a modem. Because of the high bandwidth provided by a coaxial cable, bandwidths of 30 Mbps per user can be supported, making cable one of the most versatile and converged networks. However, cable suffers from a high build cost due to the cost of digging up the road and will typically only be deployed in urban areas where the user density is high.

Digital TV   The advent of digital television will allow the delivery of broadcast data in relatively large volumes. This could speed Internet access to some of the most popular pages and provide information such as newspapers online. Digital TV systems could offer shopping malls where the contents of the shops are regularly broadcast and stored locally. The key deficiency with TV is the lack of a return channel, but for applications such as home shopping only a low-capacity channel is required, which might be supplied by the copper of the PTOs or a WLL operator. The key advantage of TV is the familiarity that people have with TV and teletext-type services.

Video distribution systems   Video distribution systems are broadly the equivalent of WLL versus copper telephony but in the world of cable systems. They attempt to provide a similar service to cable TV distribution systems but using radio. There are three key types of video distribution systems: the *microwave video distribution system* (MVDS), the *microwave multipoint distribution system* (MMDS), and the LMDS. Just like cable, these systems started as analog TV distribution systems, with the first offering being the MMDS system operating at around 2 GHz and providing around 20 TV channels. However, as time progressed, these systems also started looking for a return path to allow telephony and possibly computer data transmission. As with cable, the provision of a return path is not simple, and it is the digital variants, LMDS operating at 28 GHz in the United States and MVDS operating at 40 MHz in Europe, that will be the first to provide a return channel. However, at these high frequencies, subscriber equipment is relatively expensive, propagation is restricted, and hence it is not clear how cost effective the provision of such services will be.

LMDS provided some interest in the United States in 1998 when frequencies were auctioned across the country in the 28-GHz band. However, the money raised by the government was only around a quarter of that expected and many of the licenses have gone to startup companies where it is not clear that all will be able to find the capital required to deploy the system. Some will deploy a one-way video distribution system, whereas others will put in place a two-way broadband data system. The latter will compete directly with broadband WLL systems operating at lower frequency bands; this issue is discussed in more detail in the following sections.

Cellular and cordless radio   Cellular and cordless systems have already been discussed in Chapters 3 to 5. As access technologies for delivering fixed service, they have a number of shortcomings, including relatively poor voice quality (although this is improving all the time and is expected to be as good as wireline by 1999), lower availability, higher cost, and lower data rates. Against this is the benefit of mobility when within the house, using a single number that will reach the user wherever he or she is.

A number of operators have attempted to attack the fixed line market using cellular but generally find that they are unable to compete on price with the fixed operator. Recently, a number of interesting initiatives have been announced that may start to change this. The home base station concept within GSM allows a miniature base station to be connected to the phone socket, providing GSM coverage within the house. The wireless office concept deploys picocells within an office to provide excellent coverage and capacity coupled with mobility and a single number. If these sorts of initiatives can be provided in a cost-effective manner, then cellular systems may start to have more impact on the access market. Their key strength is that they integrate fixed and mobile communications into one terminal; this is discussed in more detail in Chapter 6.

Satellite systems   This is the most speculative of all channels. Teledesic proposed a satellite network allowing high-speed access from anywhere in the world. However, a few calculations show that this network could only handle a very small percentage of the world's traffic and will make no significant impact on other access networks.

Table 5.3 provides a summary of the key access technologies.

It is clear that WLL has a range of competitors, some that are less expensive and others that can provide higher data rates. The attraction of WLL when compared to these competitors is discussed in the following section.

5.3.3   WLL and cellular: the differences

Put simply, WLL networks are simplified cellular networks. They use the same basic architecture and the same principles of radio transmission.

Table 5.3
Comparison of the Different Access Technologies

Access Technology	Data Rates	Cost	Market Share
Ordinary telephone lines	<56 kbps	Low where already installed, high where not installed	Around 60 to 80% in developed countries
ADSL	Up to around <6 Mbps to the user, <1 Mbps to the network	Probably less than $1,000 where a copper line is already available	None at present, but may gradually gain a small share
Cable lines	Up to 30 Mbps	As with ordinary telephone lines	Varies, but typically 20 to 90% in developed countries
Mobile radio	<10 kbps	Moderate, higher than maintaining existing cables, lower than installing new ones	Rapidly gaining share, typically over 10% in developed countries
Telephony WLL	Typically <128 kbps, although broadband WLL systems are offering around 2 Mbps	Similar to mobile radio, may be slightly lower	Very little to date, but predictions that this could be more than 20% by 2005
Video distribution systems such as LMDS or MVDS	Mbps to the user, kbps to the network	Higher than WLL, but true cost not really known yet	Virtually none to date, future impact unclear

However, since the subscribers do not move, they do not need location registers, means of handing over subscribers to different cells, means of allowing subscribers to roam to different networks, and strong means for combating fraud, for example.

WLL networks do have to provide a better quality of voice than cellular networks since they are designed to replace wired networks. They need to support data transmission at the rates that can be achieved through wired networks, they need to be extremely reliable, and they need to offer communications for a lower cost than mobile radio. These requirements result in different design decisions between cellular and WLL in terms of the way the radio channels are divided and the type of antennas used, for example.

5.3.4 Technologies for WLL and LMDS/MVDS

The ideal WLL product   The ideal WLL product depends somewhat upon the environment in which it is to be deployed. For emerging countries, the ideal system would provide generally reliable voice of understandable quality at a very low price. Products based on modified analog cellular systems provide these facilities, but whether the price is low enough depends on the GDP of the country. However, in developed countries the requirements are much more stringent. To be worth deploying a WLL system must provide:

▶ Voice quality equal to the fixed network;

▶ Data rates of at least 32 kbps and preferably 64 kbps;

▶ Fax and voiceband modem support;

▶ A second telephone line.

However, the ideal system would go well beyond this to provide:

▶ Data rates equal to corporate LANs, that is, around 10 Mbps;

▶ Packet data facilities to allow efficient use of bandwidth;

▶ Bandwidth on demand and multiple lines to allow simultaneous data, Internet, voice, and video transmission.

Of course, all of these facilities should be provided at a cost lower than the revenue that can be received from the users.

Constraints on WLL systems   Technically, achieving all these requirements is relatively straightforward. Radio systems with sufficient bandwidth and processing power are perfectly possible using standard technology. The problems come when they are deployed. For a design that is technically perfect, an assignment of radio spectrum at least 10 times the bandwidth offered to a user is required for the system to be economically viable. This means that the 10-Mbps system suggested previously would need a spectrum assignment of 100 Mbps before it could be viable, and therein lies the problem. Assignments of this size are typically only available above frequencies of 10 GHz. However, at these frequencies there can be problems with rain fading; a short communications range; and because there are no standard frequency bands, manufacturers are currently unable to achieve economies of scale. In an effort to reach a compromise between cost and capacity, manufacturers have developed a wide range of systems with varying bandwidths and costs, offering operators a choice between an expensive but near-ideal system or a less expensive but less ideal system.

The current range of WLL products　There is a bewildering range of WLL products, as shown in Table 5.4, with little standardization between them. In this section some of the key systems are considered, more details can be found in [3]. WLL technologies are generally divided into those based on cordless standards, those based on cellular standards, and those custom designed for WLL. The custom category can be further divided into narrowband and broadband, with the dividing line perhaps at around 1 Mbps per user.

The progress and strengths of each of these technologies is assessed briefly in Table 5.5, noting that the products are changing rapidly and hence, the strengths and weaknesses may change, and that this overview must inevitably be somewhat superficial.

The narrowband custom technologies such as Nortel and Tadiran are now relatively well established. Both the Nortel system and the Tadiran system have been deployed in a number of commercial WLL networks, such as Ionica in the United Kingdom and networks in Colombia, while the DSC and Lucent systems have been widely trailed and were now entering commercial deployment during 1998. Equipment prices for most of these custom systems are still typically quite high, with subscriber units often costing nearly $1,000 each. There seems no reason why these prices should not fall dramatically in the next few years.

At the time of writing (1998), the broadband systems such as those from Ericsson, Floware, and Telescicom were still mostly in the test and development stage. Although there were several ongoing trials, it did not appear that any of these systems had yet entered large-scale commercial deployment. Much was unclear about these systems and the services that they offer. Propagation characteristics at the frequencies at which they operate are less well known than lower frequencies, the realistic system capacity under load was not clear, deployment rules were yet to be established, and the protocols and services to be offered were not standardized. It seems likely that trials and early deployments might take some time with these innovative new systems and that the technology will not be fully mature for a number of years to come.

Table 5.4
Comparison of Different WLL Technologies

Technology	Voice Quality	Frequency Band	Maximum Data Rate	Multiple Lines
Cordless				
DECT	Good	2 GHz	552 kbps	Up to 12
PHS	Good	2 GHz	32 kbps	In future
CT-2	Good	900 MHz	32 kbps	No
Cellular				
TACS	Poor	900 MHz	No	No
GSM	Medium	1–2 GHz	9.6 kbps	No
cdmaOne	Medium	1–2 GHz	9.6 kbps	No
Custom				
Nortel	Good	3 GHz	64 kbps	Yes
Tadiran	Good	2–3 GHz	32 kbps	Yes
DSC	Good	2–3 GHz	144 kbps	Yes
Lucent	Good	3 GHz	128 kbps	Yes
Broadband				
Ericsson AirLine	Good	10–28 GHz	2 Mbps	Yes
Floware	Good	3–10 GHz	4 Mbps	Yes
Telescicom	N/A	2–5 GHz	4 Mbps	Yes
Wireless LANs	N/A	2–10 GHz	Mbps	N/A
LMDS/ MVDS	Not known	28–40 GHz	10 Mbps to home, 10 kbps to network	No

An intriguing possibility is the use of wireless LANs as the basis for a WLL technology. Wireless LANs are designed to connect computers into office networks over a range of a few tens of meters but are relatively mature and inexpensive. By placing a directional antenna on the subscriber device and increasing the transmitter power, the technologies could be deployed for WLL systems with a range of many kilometers. Most systems will require some modifications to compensate for the increased signal delay and increased delay spread. Perhaps the greatest

Table 5.5
Strengths and Limitations of the Key WLL Technologies

Product	Strengths	Limitations
DECT	Standardized product with a relatively high data rate	Small spectrum assignment makes deployment difficult
PHS and CT-2	Standardized product	Relatively low data rate and limited spectrum
TACS	Standardized and inexpensive	Lack of spectrum, low voice quality, very low data rate
GSM and cdmaOne	Standardized and inexpensive	Lack of spectrum, moderate voice quality, very low data rate
Nortel and Tadiran	Mature and available	Relatively low data rate of around 32 kbps
DSC and Lucent	Higher data rate of around 128 kbps, ISDN compatible	Not yet widely deployed
Ericsson AirLine, Floware, Telescicom and others	Very high data rates of 2 Mbps	Generally not yet commercially ready, operates in higher frequency bands
Wireless LANs	Available and relatively inexpensive, high data rates	Not proven in a WLL deployment
LMDS/ MVDS	Provides video distribution as well as telephony	Highly asymmetric and in a very high frequency band

concern is that these systems will be relatively inefficient of spectrum and will be difficult to deploy in a cellular configuration where interference from neighboring cells might be experienced.

The final category of systems is the video distribution technologies. These are generally not classified as WLL systems since their primary role is video broadcasting rather than point-to-point communications. However, their architecture is nearly identical to WLL systems, and recently they have been acquiring return channels to enable two-way communications. These systems are known by a number of names—the early systems deployed mainly in the United States were termed MMDS and worked at 2 GHz, the digital replacement in the United States is known as LMDS and works at 28 GHz, while the European version is known as MVDS and works at 40 GHz. At present LMDS and MVDS deployments are mostly still at the trial phase. The video distribution systems are typically highly asymmetric with large downlink bandwidths of the order of hundreds of megabits per second in order to provide tens of video channels. Where return paths are provided, they tend to be limited to tens of kilobits per second and are generally intended to allow video on demand films to be selected and provide limited interaction.

Comparing reality with the ideal   The ideal WLL system could be realized by a technology based around packet transmission with a maximum bandwidth per user of 10 Mbps operating in a spectrum assignment of 100 Mbps. Some of the new broadband technologies start to approach this ideal in terms of data rates although few use packet data. However, at present, they all fall short of the ideal because they need to be deployed at high frequencies (in order to achieve a sufficient spectrum assignment), because they are still somewhat immature, and because there is little standardization of product, suggesting that the terminals might be expensive and difficult to multisource. Although the maturity will improve with time, it is difficult to see any change in the other areas.

Likely progress of WLL technologies   The wide range of rapidly changing WLL products leads to a number of questions such as:

▶ Will the existing narrowband custom-designed technology rapidly be outdated by the emerging broadband technologies?

▶ Will broadband WLL systems be based on custom-designed approaches or on wireless LANs?

▶ Will the video distribution systems develop to the extent that they become the preferred WLL technology, offering both broadband data services and video distribution?

▶ Will any standard for WLL systems emerge in the coming years to stabilize the market?

The answers to these questions are far from clear. However, it seems likely that the higher frequency of operation of the broadband systems will demand a price premium both in equipment cost and in a higher number of cells for some time to come. As a result, there will be a number of niches in the market, as shown in Table 5.6.

The (relatively) narrowband systems are already being deployed by operators wishing to compete with the PTO. These systems offer broadly equivalent services to the fixed line but at a lower price and will become increasingly ubiquitous as the equipment costs fall and the technology becomes more mature. The broadband systems will be aimed at home

Table 5.6
Possible Market Segments for WLL Technologies

System	Data Rates	Cost	Role
Modified cellular	<10 kbps	Low	Emerging countries
Narrowband custom	<500 kbps	Medium	Consumer market in developed countries
Broadband	<2 Mbps	Probably high, to be seen	Home workers and SMEs in developed countries

teleworkers and *small- and medium-sized enterprises* (SMEs) who require interconnection to LANs, intersite communications, or simply a cost-effective alternative provision of a number of voice circuits. The combination of a difficulty in gaining radio spectrum and the current immaturity of the broadband products means that such systems will probably start to be deployed during 1998 and 1999, but will not become commonplace until the year 2000 or later.

It is probably of little relevance (except for the manufacturers) whether the broadband systems are based on wireless LANs or custom-designed technologies. At present, most are custom-designed, but some wireless LAN-based products will probably emerge in direct competition during 1999.

Whether the video distribution systems will enter this market is difficult to predict. At the moment, the increased cost of operating at very high frequencies coupled with a design aimed mainly at video distribution suggests that these systems will serve yet another marketplace—that of video distribution. But as the broadband WLL products rise in frequencies toward those used by the video distribution systems, they will start to share components and may increasingly tend to be variants on the same platform.

5.4 Satellite systems for telephony

5.4.1 Introduction

For some time, cartoon characters such as Dick Tracy had wristwatch communicators that magically beamed their signals to satellites overhead so that the characters could stay in communication wherever they were. This seemed like a different world from today's mobile phones. For users within Europe, using GSM, the situation is not too bad, with relatively good coverage across Western Europe and simple roaming arrangements between different countries. However, for people in the United States, the situation is quite different. Coverage is patchy and often only exists in cities. Different standards and different companies operate across the country with the result that roaming from state to state is often impossible or at best very expensive. The U.S. standards are not widely used outside the United States, with the result that when U.S. citizens leave the country they can no longer use their cellular phones. Perhaps it is not surprising that the impetus for a global satellite system providing worldwide communications came from the United States.

The basic idea is simple. In order to cover large areas, large cells are required. Large cells need transmitters sited high up. The best position is somewhere in the sky, hence the use of satellites to provide coverage over wide areas. The implementation is much more difficult. Just launching the satellites is a high-cost and high-risk undertaking. Then there is a requirement for a global spectrum assignment, something that is rarely achieved. Finally, there is the standards minefield to negotiate. In this section the concept and progress of global satellite systems is briefly examined. More complete information can be found in [4].

5.4.2 Concept

The concept of a satellite system is relatively straightforward—the system is broadly similar to a cellular system, but with the base stations placed in space rather than located on the ground. Satellite systems are not new, and satellites have been used for fixed communications for over thirty years. These types of satellite systems typically used large satellites placed in geostationary orbits above strategic parts of the world, often across major oceans. Large Earth stations with dishes some 30m across, placed on either side of the ocean, communicate with the satellite. International fixed calls are routed by the PSTN to these Earth stations in order to pass over the satellite link to their destination. Because of the size of these dishes, such a use of satellites is clearly not appropriate for personal communications.

In the 1980s, a limited form of mobile communications via satellite became possible using the Inmarsat series of satellites. These too were geostationary; however, advances in satellite technology enabled relatively small dishes only a meter or so across to be used. Such small dishes are known as *very small aperture terminals* (VSATs). This allows shipping vessels to send signals via satellite and eventually for travelers to remote areas to take a briefcase, with an unfolding antenna, that could be aligned with the satellite to provide a voice link, or a low-speed data link. News reports from remote regions are typically still sent back using this approach.

VSATs represent the smallest and most efficient satellite terminals deployed up until 1998. However, they are a long way from the Dick Tracy ideal, or indeed from anything that a typical U.S. citizen would like to carry around with them, let alone with the need to orient the antenna. The reason why VSAT terminals have not become smaller is broadly due to the distance that the satellite is from the Earth. To be in a geostationary orbit, the satellite needs to be 36,000 km above the Earth's surface. This requires substantial transmit power and receive sensitivity and can typically only be achieved via the use of a directional antenna. It is the size and deployment of the antenna that makes VSATs inappropriate as a consumer product. VSATs also suffer from noticeable speech delay of around one-fourth of a second due to the time it takes radio waves to travel to the satellite and back; and finally, VSAT systems typically comprise less than 10 satellites, each with limited capacity.

The obvious solution to this problem is to bring the satellites closer to the Earth by placing them in low Earth orbits (LEOs) between 500 km and 1,000 km above the Earth. The shorter propagation distance reduces the delay and the power consumption to levels at which handheld terminals without highly directional antennas can easily transmit and receive signals. These lower orbits also have smaller cell sizes, and hence the total system capacity can be higher. A LEO system requires about 60 to 70 satellites for global coverage.

The concept of bringing the satellites closer to the Earth is not one that requires inspired thinking; the reasons why low orbits have not been considered before are because the satellites are not stationary in space. In a LEO, a satellite can pass overhead in less than six minutes. This requires more complex tracking of the satellites at the Earth stations and means that handover is required between different satellites. In practice, this handover is simpler than in a cellular network where the network has to deduce to which cell the mobile should be handed over. In the satellite system, the system knows which is the next satellite coming along to cover that area, so it knows exactly where the calls need to be handed. Another problem with low orbits is that the satellites do not stay in orbit for very long. They are expected to have a lifetime of only around 5 years. In order to keep a system of 70 satellites in space, that implies that 14 satellites per year will need to be launched just to replace those that are falling out of the sky.

The means whereby the call is routed to its destination, once it reaches the satellite, is worth further investigation. Some of the proposed satellite systems use "switching in space," whereas others use a "bent pipe" design. Space switching means that the first satellite to receive the call determines the destination of the call and sends the call via neighboring satellites until it is over a point on the Earth close to the destination. At this point the signal is sent down to an Earth station and interconnected with the PSTN or PLMN as appropriate. In a bent pipe system the satellite just sends back to Earth the signal it receives from the mobile. The Earth station near the mobile then uses the international fixed telephone network to route the call to the called destination. Space switching has the advantage of a reduction in cost because international calls are not required, an increase in reliability because only one network is involved, and a reduction in the number of Earth stations because it is not necessary for each satellite to be able to see an Earth station. It has the disadvantage of additional complexity in the satellite network that now has to be able to provide a switching function and call routing to neighbor satellites. The Iridium system uses switching in space. The bent pipe system has the advantage of simplicity, resulting in relatively cheap and reliable satellites, but the disadvantage that more Earth stations and international calls are required. The Globalstar system is a bent pipe system.

There is a further issue to do with economics that has recently caused the space switching systems much difficulty. Many PTOs, especially in developing countries, make large profits from international calls that are priced much higher than local calls despite the fact that the cost of providing them is only slightly greater.[3] Many of these international calls are

---

3. For a further discussion of the issues of international call costs and subsidization, see [3].

generated by visiting businessmen. These countries are concerned that the space switching systems will "steal" most of their international calls, reducing their revenue. As a result, they have decided not to allow the space switching operators access to the radio spectrum in their countries. The bent pipe systems do not have this problem, since as long as there is an Earth station in the country, the calls will be routed into the PSTN locally, enabling the international calling rates to be achieved. This problem has caused the space switching operators to deploy more base stations than is strictly required in order to cause more of the traffic to be routed on the fixed networks, removing many of the advantages of space switching.

This is an interesting lesson for the complete wireless professional. The technical arguments proved to be irrelevant in the case of a distorted pricing regime and the concerns of developing countries.

5.4.3 Economics of satellite systems

At the time of writing, in 1998, all the Iridium satellites were in place and service was expected to start later in the year. The owners of Iridium have spent an estimated $4B in putting their system into orbit and now need to recover that cost. Opinion is still divided over whether satellite systems can be profitable. Clearly the owners of Iridium have satisfied themselves that their money is invested wisely. The basic selling proposition is that users of Iridium or other satellite systems will be prepared to pay substantially more for calls made from remote areas than cellular and wireline phone users pay for their calls because there are no other options and because the users will wish to stay in contact. Iridium may charge up to $3 per minute to make a call on its phone. The question is whether there are enough people in the world who regularly travel to places where their cellular phones do not work and who would pay $3 per minute in order to keep in communications.

Iridium claims that they only need around 700,000 customers each paying an average of $3 a minute for their calls to repay the cost of the system within three years. Given that the system cost is estimated to lie between $4B and $8B, this would mean that revenues in excess of $2,000 per subscriber per year would be required. This is not an unreasonable expectation for senior management within companies. Iridium will be targeting their system very much at global executive travelers who require to stay in contact wherever they travel. With the launch of Iridium services expected in September 1998, it will not be long before it becomes clear whether this is a viable service capable of making significant revenue.

5.5 TV, radio, and other systems

TV and radio   Television and radio do not obviously fit into the category of mobile radio, although, stretching a point, the mobile reception of TV and radio is widespread. However, the complete wireless professional needs to know a little about this area because:

▶ The advent of digital television will change the manner in which TV and radio make use of radio spectrum;

▶ Digital radio will allow the transmission of data to individual users;

▶ TV and radio face the same propagation difficulties as mobile radio;

▶ Many of the technical designs used for digital TV and radio are also being proposed for cellular systems.

Current analog television systems work in a similar way to cellular systems as far as the network design is concerned. Transmitters exist in a cluster arrangement with a typical cluster size of 11. Given that transmission of an analog TV channel requires approximately 8 MHz of bandwidth, to transmit a single channel nationwide requires 88 MHz of bandwidth. Most countries typically deploy around four national channels in the UHF frequency band, between around 400 MHz and 800 MHz, typically mostly filling this band when a few channels are set aside for outside broadcast purposes and difficult coverage areas. The actual signal transmitted is a complex composite analog signal containing luminance, color, and sound information that is not discussed further here.

Television broadcasting is due to change in the next few years with the introduction of digital TV. Digital takes advantage of the compression that is possible on a TV picture in much the same manner that speech coders can compress voice signals. With TV, the current frame is almost always nearly identical to the previous frame and, hence, if the difference between frames is transmitted rather than the actual frames, this allows a dramatic reduction in the bandwidth required. However, like voice coding, the greater the compression, typically the lower the quality of the final picture. Digital TV standards allow for a range of different coding rates from around 1 to 8 Mbps. The lower rates are envisaged for lower quality critical programs such as game shows whereas the higher bit rates would be used for films and sporting events. As a result, the number of channels available to the viewer to watch could change during the day as the program content changes.

One of the key design decisions in both digital radio and digital television has been the capability to deploy *single frequency networks* (SFNs). At present, it could be argued that much spectrum is being wasted by analog TV, which broadcasts the same signal in neighboring cells but needs 11 frequencies to do so.[4] Since the same information is broadcast in each cell, in principle only one frequency should be required. However, to achieve this with analog TV transmission would require careful design of quasi-synchronous networks as discussed in Section 4.2 and even then would typically result in significant "ghosting" for viewers. With the capability of digital systems to process signals before they are placed on the screen, it is possible to deploy SFNs.

The approach adopted to achieve this has been to transmit a short pause after each bit is transmitted. This is shown in Figure 5.3.

In this figure, the first trace shows the signal from the nearest cell and the second and third traces from neighboring cells that result in signals with an increased delay (because they are further away) and a lower power. The resulting signal seen by the receiver is shown in the bottom trace. As can be seen, because all the echoes arrive before the next bit from the nearest transmitter, they do not interfere with each other. Indeed, by adding all the signals from each of the transmitters a stronger overall signal is received, one that is less vulnerable to interference because of the diversity of reception (from a number of transmitters). However, the down side is obvious, the bandwidth required by the original signal has increased significantly as a result of sending shorter pulses. In the preceding example, where the pulse width was only one-fourth

---

4. This is not the case in the United States, where typically cities are sufficiently far apart that the same frequencies can be deployed in each city without any interference. In Europe, the population is more evenly spread and better geographical coverage is required with the result that interference can become problematic.

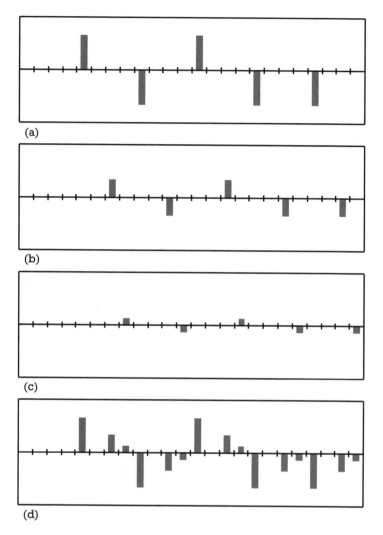

Figure 5.3 The signal received in an SFN: (a) the signal from the closest transmitter, (b) the signal from the second closest transmitter, (c) the signal from the third closest transmitter, and (d) the composite signal as seen by the receiver.

of the total possible pulse width, the bandwidth requirement would be increased by a factor of 4. Nevertheless, if this removed the need for a cluster size of 11, the net gain in spectrum efficiency would be 11/4 = 2.75 times.

Some simple calculations reveal further problems with this arrangement. If it is assumed that interfering base stations might be up to, say, 50 km away, then the delay on receiving a signal from these base stations will be around 0.2 ms. If there needs to be a delay of 0.2 ms after the transmission of each pulse, then even if the pulse length is negligible this limits the data rate to 6k baud. Even with the proposed 16- and 64-level quadrature modulation, allowing four or six bits per symbol, respectively, this still only allows a maximum data rate of around 24 or 36 kbps, far too low for the transmission of digital video.

The solution to this problem has been to split the data to be transmitted into a large number of streams and to encode and transmit each of these streams separately. Using up to 8,000 separate streams, the data rate on each stream can be brought well within the limits calculated previously. Such an approach is well known from mobile radio research as *orthogonal frequency division multiplexing* (OFDM), even if it has never been deployed in a commercial mobile radio system. In a practical arrangement, OFDM works by passing the incoming data through a Fourier transform (which has the same effect as splitting it into a number of streams), transmitting the resulting data, and then performing an inverse Fourier transform on the received data in order to arrive back at the original information. OFDM is a complex topic that is discussed in more detail in [5]. In summary, the key differences between digital radio and TV transmission and cellular are the use of OFDM and the use of QAM.

Mobile radio systems do not need OFDM because there is no possibility of an SFN since different information is transmitted in each cell. They tend not to use quadrature modulation because the quality of the mobile radio channel is so poor that it is not possible to accurately distinguish between a number of different modulation levels. For a more detailed discussion of the issues surrounding the use of QAM, see [5].

Despite the fact that, in principle, the introduction of digital TV should result in a dramatic improvement in the spectrum efficiency and hence the possibility that some spectrum could be returned to the regulator and then possibly assigned to cellular operations, it is still far from clear whether this will be the case. The arguments are complex and beyond the scope of the discussions here, but some of the key points include:

▸ More spectrum than at present will be needed during the transition from analog to digital when simultaneous broadcasting on both formats will be required and this transition could take as long as 15 years.

‣ Digital TV will need to offer many more channels than the current analog TV in order to persuade viewers to buy a new TV set.

‣ Digital TV enables new facilities such as home shopping, which may require additional bandwidth.

‣ SFNs may not be suitable for all channels because they do not allow regional variations such as local news and different advertisements in different regions as are currently transmitted.

Other users of the radio spectrum   There are many other users of the radio spectrum with which the complete wireless professional only needs a passing familiarity.

‣ Aeronautical radio and radar for civil aviation purposes—in particular, the aeronautical radar and instrument landing systems take up a large amount of spectrum below 2 GHz;

‣ Military usage, which tends to be confidential but uses a large percentage of the spectrum under 2 GHz;

‣ Fixed link usage (fixed links are discussed in Section 7.1.5), which can be used to support mobile radio networks; fixed link bands extend from 2 GHz upward;[5]

‣ Amateur and research users, which have small assignments throughout the mobile spectrum band;

‣ *Industrial, scientific, and medical* (ISM) uses, which include, for example, the use of radio to dry leather hides and heat certain products; this band is unlicensed and recently a number of WLL operators started using the band for communications, using frequency hopping and CDMA technology to overcome the unknown and time-varying interference.

---

5. In some cases, mobile radio engineers may have to interwork with fixed links. For example, in the U.S. PCS bands, the spectrum was sold to the mobile radio operators with fixed links already in the bands. The operators had to decide whether they would work around the fixed links or pay the fixed links operators to move to a different band. This legacy use of the band continues to cause interference problems.

For more details about the use of the radio spectrum in any particular country, the national spectrum manager can normally provide the required information. For example, complete details of all spectrum allocations in the United Kingdom can be found from the U.K. Radiocommunications Agency [6] while details of U.S. allocations can be found from the FCC [7].

References

[1]   Tuttlebee, W., ed., *Cordless Telecommunications Worldwide*, Berlin: Springer-Verlag, 1997.

[2]   Phillips, J. A., and G. Mac Namee, *Personal Wireless Communications with DECT and PWT*, Norwood, MA: Artech House, 1998.

[3]   Webb, W., *Introduction to Wireless Local Loop*, Norwood, MA: Artech House, 1998.

[4]   Jamalipour, A., *Low Earth Orbital Satellites for Personal Communication Networks*, Norwood, MA: Artech House, 1997.

[5]   Webb, W., and L. Hanzo, *Modern Quadrature Amplitude Modulation*, New York: John Wiley, 1994.

[6]   Information can be found at
      http://www.open.gov.uk/radiocom/rahome.htm, which also
      contains links to most other spectrum managers around the world.

[7]   The FCC website can be found at http://www.fcc.gov.

CHAPTER

6

Interfacing with fixed networks

Co-existence: what the farmer does with the turkey—until Thanksgiving.

John Berry

6.1 The need for fixed networks

The complete wireless professional needs to know all about fixed networks. At a minimum, the mobile radio network will typically need to be connected to a fixed network at some point. Further, interconnection of different parts of the mobile radio system may be via fixed networks. Fixed networks are complex topics and a good introduction to them can be found in [1, 2].

Important points to understand about fixed networks are:

▶ The outline architecture of a fixed network;

▶ The concept of intelligent networks since future cellular system design will need to incorporate intelligent network philosophies;

▶ How numbering and addressing schemes are devised since mobile radio systems need to be part of these;

▶ The basic interconnection mechanism using E1 or T1 links;[1]

▶ The many different protocols that are in use on the fixed networks and the most appropriate protocols to use in any particular situation;

▶ The convergence of fixed and mobile networks expected to occur over the coming years.

6.2 Fixed network architectures

Outline   In outline, a fixed network consists of one or more switches, a number of distribution points, and wires to each of the subscribers' premises. The distribution points are used to gather a number of subscriber connections into a larger bundle of cables that can then be more easily routed back to the switch. Each switch is likely to be connected to a number of other switches using diverse routing that allows an alternative to be found in the case of failure of a single link.

Modern networks are constructed either on a circuit-switched or a packet-switched principle. In circuit switching, a dedicated circuit is established between the two users for the duration of the call. In a packet-switched system, packets are sent whenever there is information to be conveyed and these are routed along a shared resource by the network. The PSTN is a good example of a circuit-switched network, whereas the Internet is a good example of a packet-switched network. Both will be discussed in this section.

Switch design   Switch design is complex and a detailed description is beyond the scope of this book. Simplistically, in the case of circuit switching, the switch works by overlaying all the incoming lines over all

---

1. The decision as to whether to use the T1 or E1 standard is typically made on a regional basis, with T1 being used in the Americas and E1 in Europe.

the outgoing lines in a grid structure and pulling together the lines at the crossover point when a connection is required. However, such a simple switch rapidly becomes unmanageable as the number of cross-points increases by the square of the number of incoming lines. Switches were only designed in this manner until around the middle of this century. The next generation of switches used the principle of fan-in and fan-out where the incoming lines were switched onto a smaller number of internal lines when a call was taking place. This smaller number of internal lines was then switched across the outgoing lines. For example, in the case of 10,000 incoming and outgoing lines, in a simple exchange 100,000,000 cross-points would be required. For 10,000 incoming lines with a typical traffic level the Erlang formula predicts that about 560 lines will be required to produce acceptably low levels of blocking. At the fan-in point the 10,000 lines cross with the 560 lines, and at the fan-out point the same number of cross-points are required. The total number of cross-points needed is $2 \times 560 \times 10,000 = 11,200,000$, only around 10% of the number of cross-points required in the first case. All these types of switching are referred to as space switching because a switch taking up a certain amount of space connects and disconnects physical resources.

Modern switches use digital switching in a time-space-time format. This takes into account the fact that the incoming lines are typically E1s on which the calls are encoded using TDMA. A modern switch needs to be able to take the incoming signal on, say, timeslot 6 of a particular E1 link and to pass it to, say, timeslot 18 of a particular outgoing link. Hence the need to switch first in time, second in space, and third back in time again. A simplified diagram of a time-space-time switch is shown in Figure 6.1.

In this switch, an incoming call on timeslot 3 of incoming line 1 is first switched in time to timeslot 30. Then the entire E1 on the left side of the switch is mapped onto the E1 on the right of the switch corresponding to outgoing line 2. Timeslot 30 is then mapped onto the final timeslot used, timeslot 1, on the outgoing line. In practice, such an arrangement would not work since it might be required to switch timeslot 3 onto outgoing line 2 but timeslot 4 onto outgoing line 1. Hence, the space switching part can switch 32 times within a frame so that incoming line 1 is connected to outgoing line 2 during the 8 bits corresponding to timeslot 3 but then connected to outgoing line 1 during timeslot 4. Such a switch has a similar advantage to the fan-in fan-out switch in that the number of cross-points are significantly reduced. Most modern digital switches are designed along this principle.

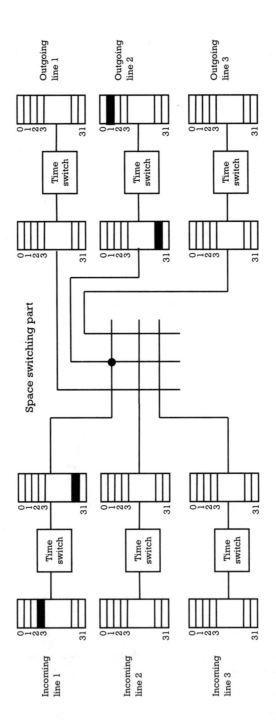

Figure 6.1 A simplified time-space-time switch.

In the case of a packet switch, a somewhat different arrangement is required. A packet switch receives incoming packets and stores them in an input buffer. It then reads packets out of the buffer, examining the address on the front of each packet. Based on the address, the packet is then sent out on the most appropriate output line. Because of the ability to tolerate a short delay, and hence the need to process only one packet at any particular point in time, packet switches can be considerably simpler than circuit switches.

Intelligent networks Modern fixed networks are constructed around an *intelligent network* (IN) philosophy, and increasingly mobile radio networks are also starting to be designed around the same philosophy. INs are intended to allow the easy introduction of new services without the need to replace or upgrade switches. Essentially, an IN separates out the hardware and the software in a network, allowing new software to be written when new services are required. To work effectively, INs need a standardized software language, a basic "operating system," and a means of introducing new software onto the network. The IN standards cover these areas. The benefits of intelligent networks are intended to be:

▶ The rapid introduction of new services with minimal impact on the existing architecture through the downloading of new service script software;

▶ The reduced cost of introducing and enhancing new services;

▶ The ability to rapidly reconfigure services to meet market needs;

▶ The ability to give limited control to selected customers so they can make authorized changes specific to their own network—an example of this might be a company subscribing to a centrex service and making some changes to the manner in which certain numbers are processed.

The conceptual model of the IN as envisaged by current standards bodies has four planes, each representing different levels of IN abstraction as shown diagrammatically in Figure 6.2.

The *service plane* represents a service-oriented view of the IN and does not contain any information regarding the implementation of the network services. This would contain a list of all the services available, along with an associated list of all the functions that they must call. For

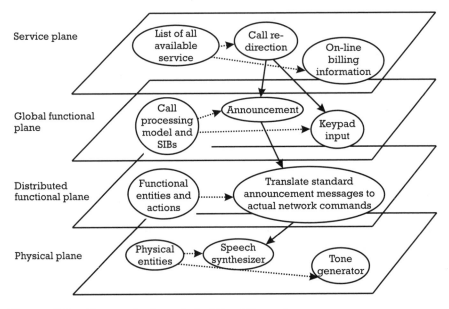

Figure 6.2　Conceptual model of the IN services.

example, there would be an entry for call redirection alongside its access code, say, *50. Such a function might require an announcement and a call translation.

The *global functional plane* resides below the service plane and provides a view of different functional entities of an IN network. The plane contains the call-processing model and *service-independent building blocks* (SIBs). Example SIBs are: announcement, a feature that may be required by a wide range of services; information input through the numeric keypad, again a widely used function; and charging procedures, for example. The SIBs are defined independently of any physical consideration. SIBs are used as building blocks in the creation of services.

The *distributed functional plane* resides below the global functional plane and is used to define the *functional entities* (FEs), their actions (FEAs), and their relationships and information flows. The functional entities translate the non-network specific SIBs into physical commands that can be sent into the network to accomplish tasks. It is within this plane that the details of alternative networks start to differ.

Finally, the *physical plane* identifies the different types of physical entities and their information flows. It contains all the peripherals and control nodes that form the structure of the network and are used to carry and process the information.

A functional model of an IN is given in Figure 6.3. Calls are routed from user to user through the *service switching point* (SSP), effectively the standard network switch upgraded to IN functionality. This unit has the capability to detect a requirement, or trigger, for an IN service. Requests for IN functions are then formulated and passed either to the *service control point* (SCP), services node, or adjunct, depending on network implementation. These requests are sent using CCITT7 signaling using special parts designed for IN called the *signaling connection and control part* (SCCP), the *transaction capability application part* (TCAP), and the *intelligent network*

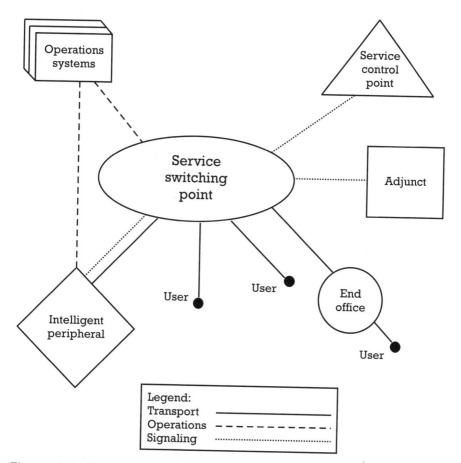

Figure 6.3 Functional model of an IN.

application part (INAP). SCPs are connected through *common channel signaling* (CCS) networks that allow a number of SCPs to be interconnected. This is used when accessing centralized databases or for person locator services. Interaction with users is provided through the *intelligent peripheral* (IP), which can issue voice messages to the user and accept key pad or conceivably voice inputs. The IP receives instructions from the SSP and relays received responses back to the SSP.

For mobile communications, all the functions of the IN described previously are still applicable. In addition, there may be a need for IN functions to allow mobility and call management. These include location information retrieval and updating on call routing, handover, charging, and maintenance. These additional service features could result in dramatic increases in service traffic, especially as cell sizes reduce and mobile densities increase, causing frequent location updating for a large number of users. Current mobile systems incorporate the functions mentioned previously into their own pseudointelligent networks, leading to the possibility of two INs, one in the mobile system and one in the PSTN to which it is connected. It is the view of the fixed link community that these should be integrated into one IN in the future. For this to happen, more integration between fixed and mobile standards authorities will be required.

Telecommunications management network   The *telecommunications management network* (TMN) is intended to be a means to allow the effective management of networks. Network management is discussed for mobile radio networks in Section 9.2. TMN is the equivalent for the fixed network. The basic problem with existing fixed networks has been that the various parts of the network often come from different suppliers; for example, the switch may be manufactured by someone different from the *synchronous digital hierarchy* (SDH) link. Each manufacturer will provide a network management system for their part of the network, but these will typically not interwork. The result is a proliferation of different management systems; and because they do not interwork, a fault in one part of the network can result in alarms appearing on a number of different network management systems, taking the manager of the network some time to determine where the fault actually lies. TMN presents a standardized interface to which all the manufacturers can pass information concerning their part of the network. This allows a single network management system to be used. TMN relies on an interface defined as the Q3 interface, the *common management information protocol* (CMIP), and *managed objects* (MOs).

Numbering and addressing plans   All telephones, including mobile phones, need a unique number. The administration of a numbering plan is important for the continued operation of the world's telephone networks. The most relevant addressing scheme for mobile radio is the ITU-T E.164 recommendation. This recommends the 00 international prefix, the 0 prefix for trunked calls, and a maximum length, excluding international prefix of 15 digits. The structure of an E.164 number is shown in Figure 6.4.

E.164 allows a number of key features.

▶ Direct dial in allows the last few digits at the end of an ISDN number to be transferred to a customer's PABX, enabling the call to be further routed inside the company without a switchboard operator.

▶ Subaddressing allows up to 40 additional digits to be added on top of the ISDN number to allow further routing on the customer premises.

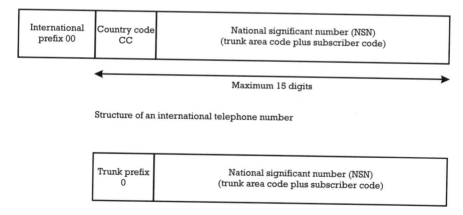

International prefix 00	Country code CC	National significant number (NSN) (trunk area code plus subscriber code)

Maximum 15 digits

Structure of an international telephone number

Trunk prefix 0	National significant number (NSN) (trunk area code plus subscriber code)

Structure of a national telephone number

Figure 6.4   Structure of national and international telephone numbers.

6.3 Fixed network protocols

The mobile radio engineer will not typically be concerned with the protocols that are used between the fixed phone and the switch because this is a topic for the fixed telecommunications engineer. However, a knowledge of the protocols running between the switches is important since this is the manner in which the mobile radio network will be interconnected into the PSTN. This section provides an overview of the protocols in use.

The basic framing structure   On fixed lines, voice is typically carried at 64 kbps using PCM coding. Lines between switches are used to carry a number of voice channels. The European standardized rates for these links are:

2.048 Mbps	termed an E1 link
8.448 Mbps	termed an E2 link
34.368 Mbps	termed an E3 link
139.264 Mbps	termed an E4 link
564.992 Mbps	termed an E5 link

In the United States there is a different nomenclature:

1.544 Mbps	termed a T1 or DS1 link
6.312 Mbps	termed a T2 or DS2 link
44.736 Mbps	termed a T3 or DS3 link
139.264 Mbps	termed a DS4 link

These two systems are incompatible at all levels, but protocol converters now exist that allow one system to be connected to the other. However, in the connection process, channels typically have to be wasted; for example, when connecting an E1 to a T1, two T1s are required, providing a total bandwidth of 3 Mbps, of which 1 Mbps remains unused.

Higher rate links are assembled by multiplexing from lower rate links. For example, an E2 is obtained by multiplexing four E1s together and an E3 by multiplexing four E2s. Hence, for the higher rate links, there may be a significant hierarchy of multiplexers and demultiplexers required to extricate the data for one particular user. For the remainder of this chapter, the European E1 system is used for further discussion. However, the U.S. system is very similar in design, and most of the points apply equally to the U.S. system.

The E1 carrier has a strict framing format. It consists of a framing structure, similar to the TDMA framing structure of GSM, as described in Chapter 3, which repeats every $125\,\mu s$. This frame is divided in 32 timeslots marked timeslots 0 to 31. Within each timeslot, eight bits from each user are transmitted. Of these 32 timeslots, only 30 can be used to carry user data. Timeslot 0 contains a flag used to signal the start of a new frame while timeslot 15 is used for signaling purposes. The signaling information is used to set up a call on any of the other timeslots within the E1 and will contain information about the destination of the call and the timeslot on which it will be carried. Timeslot 0 is also used for synchronization purposes. The two switches interconnected by the E1 link will often be using different clocks. Although every effort is made for the clocks to be running at the same speed, they are not actually synchronized and are often termed plesiochronous to indicate that they are unsynchronized but adjusted to run as closely to each other as possible. In the case that the clock in the transmitting exchange is running faster than the clock in the receiving exchange, more bits are sent than can be received. To overcome this, the receiving exchange can simply remove some of the bits in timeslot 0. In the opposite case, the receiving exchange can add some extra bits into timeslot 0 to make up the rate. An E1 system of this form is said to belong to the *plesiochronous digital hierarchy* (PDH).

More modern systems are designed using fully synchronous networks and are referred to as being part of the SDH or *synchronous optical network* (SONET). In these systems all the different exchanges must use the same clock. Typically, in any country there will be a master clock. This master is transmitted via wire to each of the exchanges directly below it in the hierarchy. Each of these exchanges then further distributes the clock. They also typically have their own clock that is tied to the master clock but, in the case of loss of master clock, can continue to run so that the telephone system can continue to work. Different countries can choose to link their master clocks if required so that they are synchronized. More recently, another option has been to use the timing reference transmitted by the GPS as the clock source at each of the exchanges. SDH has a major advantage over PDH in that single channels can be inserted into and removed from the stream without the need to demultiplex back to an E1 prior to removal or insertion. Demultiplexing is necessary in PDH in order to remove the additional framing bits according to the difference in clock rates at each of the various level but is not required in SDH. Because of this advantage, most modern systems are synchronous.

Interexchange signaling It was mentioned previously that timeslot 15 on the E1 link is reserved for signaling associated with setting up calls. An agreed protocol is required for this signaling such that different exchanges can interwork. There are a number of different protocols that can be used for this purpose. Some of the key ones are:

 ▶ CCITT6: A common channel signaling system for use between analog exchanges with a signaling speed of 2.4 kbps;

 ▶ CCITT7: A complex and multipurpose common channel signaling system for use between digital exchanges with a signaling speed of 64 kbps;

 ▶ CCITT R1: A regional signaling system formally used for trunk signaling in North America;

 ▶ CCITT R2: A regional signaling system used within Europe.

Some of the key signaling protocols are described in more detail.

The R2 signaling system R2 is a form of channel-associated signaling that means all the signaling that is relevant to a particular channel passes down that channel itself. This differs from the E1 arrangement described previously where there is a separate signaling channel for the 30 connections carried on the link. Before digital exchanges were widespread, associated channel signaling was the only possible way of transmitting signaling information, but it has the disadvantage that a separate signaling "reader" is required on every single line, whereas with common channel signaling the number of readers is considerably reduced (by a factor of 30 in the case of the E1 arrangement)—hence, the trend in modern systems toward common channel signaling. R2 line signaling is carried out of band in tones at 3,825 Hz that are filtered out in the receiving device that only allows voice signals in the band 300 Hz to 3.4 kHz to pass. By being out of band, the signaling does not disturb the voice traffic but is restricted to a low data rate. The signaling used is relatively simple, moving a state machine through six different states as shown in Table 6.1.

R2 works by sending a signal (setting the tone on or off) until a signal is received from the other end. The received signal both acts as a

Table 6.1
The Use of Tone Signaling in R2

State	Outgoing Tone	Incoming Tone	Meaning	Moves to State
1	On	On	Circuit idle	2 or 6
2	Off	On	Seized	3 or 5
3	Off	Off	Answered	4 or 5
4	Off	On	Clear	3 or 5
5	On	Any state	Release	1
6	On	Off	Blocked	1

confirmation and allows the state machine to be moved to the next state. This acknowledgment process is known as compelled signaling.

There is also a multiple frequency version of R2 where the signaling is carried in band during the set up of the call and can be used to carry the dialed number. This signaling system is known as R2 MFC. R2 is only of interest when needing to interface with older exchanges; modern digital exchanges will typically make use of CCITT7 signaling.

Integrated services digital network   The *integrated services digital network* (ISDN) is a network providing end-to-end digital connectivity to the user [3], unlike most fixed connections at the moment that are analog and require modems to transmit digital data. ISDN provides the services of voice, 64-kbps data, facsimile, telephony, and video-telephony as well as a range of supplementary services. ISDN standards are set worldwide by the telecommunications division of the *International Telecommunications Union* (ITU-T) and contain two types of customer-to-network interfaces: the *basic rate access* (BRA) and the *primary rate access* (PRA). Both consist of a number of bearer (B) channels at 64 kbps and a data (D) channel at either 16 or 64 kbps, which is not actually used to carry user data but only to carry signaling relating to the call. BRA consists of 2B + D channels, using a 16-kbps D channel; whereas PRA consists of 30B + D channels, using a 64-kbps D channel. In order to install BRA ISDN at a user's premises, the bandwidth of the subscriber connection must be increased to allow 144 kbps. This is achieved by removing loading coils on the circuit but may not be possible for all installed lines; hence, it may not be possible to universally install ISDN to all subscribers. PRA, at 2 Mbps typically requires a special connection, although with the advent of new digital subscriber line technologies (see Section 5.3.2), it is becoming increasingly possible to carry these rates across some standard telephone lines. ISDN has the advantage that a BRA can be used to send two simultaneous messages; for example, one of the B channels could be used for voice and the other for data. However, when only one line is in use, the two available channels can be combined to achieve a higher data rate than would have been possible on one channel alone.

In order to support ISDN a digital switch is required. The switch must use 64-kbps digital voice channel switching and support common channel signaling between exchanges. Typically, the interswitch protocol adopted is CCITT7 along with the specific part of the standard required for ISDN, the *integrated services user part* (ISUP).

Signaling system No. 7   CCITT7 is also widely known as SS7 (and C7). Most digital telephone networks around the world make use of SS7. It is essentially a common channel signaling system that enables two digital exchanges to communicate directly and interact in a manner that allows digital transmission. SS7 is designed in a layered and modular manner and can support a range of applications including:

▶ ISDN;

▶ Intelligent networks;

▶ Mobile services;

▶ Network operation and management.

New user parts can be written to support new applications (e.g., during the standardization of GSM, the *mobile application part* (MAP) was written to allow MSCs to interconnect using SS7). The existing user parts include:

MTP	message transfer part
SCCP	signaling connection and control part
TUP	telephone user part
DUP	data user part
ISUP	integrated services user part
TCAP	transaction capabilities application part
OMAP	operations and maintenance application part
INAP	intelligent network application part
MAP	mobile application part

The MTP and the SCCP are the basic parts of the system allowing message sending. The other parts tend to use MTP and SCCP to carry their higher level messages. Between two SS7 switches there will be a signaling network and a traffic carrying network. In most cases these will be integrated within the same E1 connection, but this need not be the case and the signaling network can be quite separate from the traffic network.

Worthy of a short mention is MAP, the signaling system in use by newer digital mobile radio networks. It is used between a mobile telephone network exchange (normally known as an MSC) and an intelligent network database (the HLR, VLR, or EIR). MAP allows interrogation of these databases and responses returned in an appropriate format.

X25   X25 is a protocol for packet switching networks and controls the manner in which computers producing packet information interwork with the packet network. Some aspects of X25 are used within the GSM network when sending signaling information related to, for example, call setup between the switch and the BSS. X25 is a connection-oriented protocol. At the start of a session a virtual circuit is established before data are transmitted. Packets of data may be anything between 3 and 4,100 bytes, allowing up to 4,096 ASCII characters to be sent in one packet. A basic packet consists of an 8-bit flag (with bit pattern 01111110), an 8-bit address, 8 bits of control information, the user data, and a 16-bit frame check sequence that detects whether there have been errors in the received data. The address field is only used to determine whether the information is a command or a response, the address details of the user already having been established at the time that the virtual channel was set up. X25 is probably the most widely used packet switching protocol to date.

The strength of X25 networks is their high reliability in transporting data accurately across lines even with high error rates, whereas their weaknesses lie in the high delays that can be incurred on data transmission, which is becoming increasingly problematic on computer data exchange.

Frame relay   Frame relay has come about due to the shortcomings in X25, as mentioned previously. X25 was developed at a time when signaling rates were around 9.6 kbps and errors on lines were frequent. X25 is less suited for today's high-speed, high-reliability lines. Packet transmission on high-quality lines does not require the acknowledgment and windowing provided by X25. Frame relay leaves the task of error correction and acknowledgment to the higher layers, concentrating only on rapid transfer of the raw information. Because it does not need to wait for acknowledgment or check for errors, transmission can be much more rapid. Frame relay is well suited to data rates of between 64 kbps and 2 Mbps and in applications such as LANs and wide-area computer networks. Above 2 Mbps, ATM is more appropriate. Frame relay includes network management to cope with information overload in the cases that there is a sudden increase in the information being sent from the computers linked to the network. Frame relay makes use of a *committed information rate* (CIR) that is agreed upon between the network and the users at the time that the connection is set up and guarantees a minimum rate. The network also agrees on an *excess information rate* (EIR), higher than the CIR, at which information can be transmitted when there is capacity on the network. As the network becomes heavily loaded, *explicit congestion messages* (ECMs) are sent to the users, who should reduce their transmission rate to the CIR.

The frame format for frame relay is very similar to the format used in X25. There is a flag, address, control field, information field, and frame check sequence. The information field can be up to 65,536 bytes in length.

TCP/IP   The *transport control protocol/Internet protocol* (TCP/IP) [4] was developed from the UNIX-based computers used by government and academic institutions at the time that the Internet started to emerge. Unlike the other packet protocols discussed briefly here, TCP/IP is a connectionless protocol, so each packet is treated as a completely separate entity by the network. The Internet protocol is the equivalent of an OSI layer 3 protocol, where the 7-layer OSI stack is shown in Figure 6.5, while TCP provides the layer 4 functionality. An Internet data message is quite complex. The framing format includes the header length, the type of service, the total length, the identification, flags, a field describing how long the message should be held before being deleted in the case of delivery being problematic, protocol details, a header checksum, source and destination addresses, options padding, and finally the user data. Addresses are provided as 32-bit information in the form 123.234.200.134, where the four numbers can each have values up to 255.

Asynchronous transfer mode   ATM is a connection-oriented packet switching protocol designed for the rapid transmission of data in broadband networks [5, 6]. ATM was designed to allow:

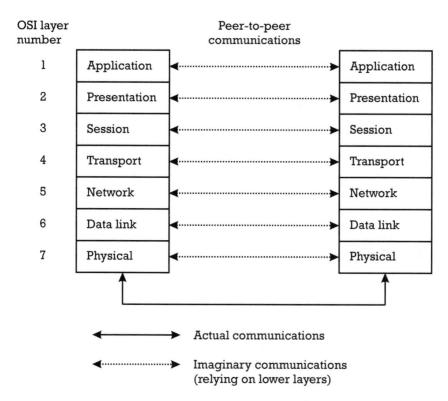

OSI layer number

Peer-to-peer communications

1	Application ◄···►	Application
2	Presentation ◄···►	Presentation
3	Session ◄···►	Session
4	Transport ◄···►	Transport
5	Network ◄···►	Network
6	Data link ◄···►	Data link
7	Physical ◄···►	Physical

◄───────────► Actual communications

◄·····················► Imaginary communications
(relying on lower layers)

Figure 6.5 OSI standard layered model of a telecommunications network.

▶ The simultaneous transmission of different telecommunications services;

▶ A contract established at connection setup that allows the end user to specify the required bit rate and connection quality including details as to the maximum permissible delay;

▶ The ability of the network to determine when new connections cannot be granted because the load on the network is such that contracts might not be met if new loads were accepted;

▶ The ability to carry delay-sensitive voice and video traffic with a minimal delay;

▶ A high efficiency in the use of bandwidth.

Unlike the other packet protocols discussed, ATM uses packets of a fixed size. These contain a 5-byte header and 48 bytes of user data. The small size of the packets requires that most information is broken up for transmission but ensures that high-priority frames do not have to wait for the passage of a large frame before they can be transmitted, thus allowing guaranteed low delays for some traffic. ATM is the most appropriate packet protocol to use for transmission rates above 2 Mbps. Because of its low delay, it can carry information that would normally have been passed down a circuit-switched protocol and so can replace both packet- and circuit-switched protocols on high speed lines.

6.4   Fixed mobile convergence

6.4.1   Introduction

Fixed mobile convergence (FMC) is a much talked about, although often confused, topic, relating to the concept that instead of fixed and mobile networks being separate entities, as is the case now, they will converge over the coming years. This section provides an overview of the possible means by which FMC might be achieved, their advantages and disadvantages, and their likely progress over the coming years. The section starts with a definition of fixed mobile convergence and then moves on to examine the currently identified solutions of:

 ▶ Network overlay solutions to give the appearance of FMC;

 ▶ Network integration solutions that put some of the mobile and fixed functionality on a joint platform;

 ▶ Solutions that integrate cordless and mobile technologies to provide a single phone;

 ▶ Substitution solutions that attempt to carry all the traffic on the mobile network.

The section then concludes with some predictions as to how the FMC market will develop over the coming years.

6.4.2 Defining fixed-mobile convergence

FMC is the interworking of fixed and mobile telecommunications. At present, FMC is mostly concerned with voice, and the simplest FMC concept is that an individual has a single phone number that people can dial and result in his or her fixed or mobile phone ringing as is appropriate. This concept is readily extended to fax and data calls, with these calls being routed to appropriate computers or fax machines depending on the location of the called party. Finally, the convergence should allow for there to be one voicemail box that can be accessed from both the fixed and the mobile phone.

Complexity arises from the use of FMC by different parties. For the fixed provider, FMC is a tool by which they can prevent traffic migrating onto mobile networks by providing enhanced functionality that will encourage the user to use the fixed phone whenever possible. For mobile providers, FMC is a tool they can use to steal traffic from fixed networks by offering the equivalent of a fixed service on the mobile phones. For the fixed provider, FMC is the bringing together of fixed and mobile communications on the same platform, whereas for mobile providers it is the avoidance of making or receiving calls on the fixed platform.

Despite this difficulty, the overriding concept is that the functionality that the user experiences should be as if they only had one phone, regardless of how this is achieved.

6.4.3 Possible solutions

There are a number of different possible solutions to FMC. In this section, the following four solutions are examined that cover the range of different possible means to achieve FMC:

- Network overlay solutions to give the appearance of FMC;

- Network integration solutions that put some of the mobile and fixed functionality on a joint platform;

- Solutions that integrate cordless and mobile technologies to provide a single phone;

▶ Substitution solutions that attempt to carry all the traffic on the mobile network.

In many cases, these solutions can be used in combination as well as in isolation; for example, network overlay solutions can be coupled with the network integration solutions.

6.4.3.1 Network overlay solutions—the personal assistant

Definition   The overlay solution or personal assistant is an IN function that resides typically in the fixed network. A user has a single number that directs the call to the personal assistant. Once it arrives there, it can be redirected to the fixed line, mobile, voicemail box, fax machine, or appropriate computer. The redirection is typically performed based on information provided by the user such as their likely movements and a knowledge of whether the mobile is contactable. In some cases, the call is first directed to the fixed line and, if not answered, is redirected to the mobile. A very simple way to achieve almost the same functionality is to forward the fixed phone to the mobile phone when leaving the office.

▶ *Cost:* This capability is relatively inexpensive to provide, requiring only an IN platform. In the case of calls that terminate on the fixed network, virtually no additional routing is required. For calls that terminate on the mobile network, some additional routing into the mobile network may be required.

▶ *Call billing:* Since it is not known to the caller whether the call will terminate on a fixed or mobile phone, either all calls to personal assistants must be made more expensive or the called party must pay an additional fee for incoming calls that were redirected to the mobile phone. In the case of call-forwarding from the fixed to the mobile phone, the called party pays for a mobile call for all incoming calls.

▶ *Supplier capability:* This service can either be provided by the fixed network operator or a third party supplier of value-added services. In a liberalized environment there is the possibility of competition in the provision of personal assistants.

▶ *Revenue potential for operator:* In principle, the personal assistance might result in fewer calls, since fewer messages are now left on voicemail. However, by charging calls at a premium rate there is potential for the service provider to make increased revenue.

▶ *User needs:* This service provides the primary requirement of a single telephone number. However, the user needs to interact with the call redirection system to ensure that calls are directed appropriately depending on time of day and location. There is typically a delay in call setup that can be annoying for callers while the assistant determines the type of call and appropriate destination.

▶ *Regulatory barriers:* There are few regulatory barriers. It is helpful to have the ready availability of personal numbers and to have a liberalized environment where service providers can interact with the fixed network operator in able to ensure competition.

Advantages This service can be provided with little change to the network and at a relatively low cost. It meets the fundamental requirement of FMC and can be provided in a competitive manner.

Disadvantages The service is typically more costly than the status quo. The need for intelligent call redirection can be tiresome for the user and result in call setup delays.

6.4.3.2 Network integration solutions—joint platforms

Definition An operator who owns both a fixed and a mobile network integrates as many of the functions of the networks as possible onto joint platforms. Typically, the first function that is integrated is billing, resulting in a single bill for both fixed and mobile usage. Following from this, voicemail and IN services can be integrated. Finally, it may be possible to integrate some of the operations and maintenance and switch functionality. Since this form of FMC does not allow for a single number for fixed and mobile services, it does not strictly meet the definition of an FMC service. However, it can, of course, be linked with an overlay-type IN solution providing the functionality described previously.

▶ *Cost:* The cost to the operator is in the procurement of platforms able to handle fixed and mobile services. However, in the long term the consolidation of platforms is likely to result in a cost reduction for the operator.

❱ *Call billing:* Billing for incoming calls is unaffected since this form of FMC does not provide a single number.

❱ *Supplier capability:* This can only be provided by the PTO in a country and only then if they also have a mobile network. Since only a subset of the mobile subscribers in the country will be using that particular mobile network, this will only be available to a minority of the population.

❱ *Revenue potential for operator:* There is little potential for increased revenue. The only possibility is reduced churn from the mobile users as a result of the enhanced service that they are being provided.

❱ *User needs:* Few of the user's needs are met. The only distinguishing feature for many users is the delivery of a single bill. This is generally thought to provide little added value.

❱ *Regulatory barriers:* There are no barriers for the PTO. Unless there is liberalization that has resulted in a significant competitor to the PTO, there will be no other company able to provide the service.

Advantages   This provides potential cost savings for the operator and the advantage of a single bill for the user.

Disadvantages   This solution does not provide a single number and, hence, does not meet the requirements of FMC.

6.4.3.3   Advanced network convergence—DECT/GSM integration

Definition   This is a solution proposed by BT under the name "OnePhone." The concept is that DECT cordless base stations are installed within the office and are connected to the existing fixed lines into the office. The DECT system is then connected into the GSM switch of a mobile operator so that location information relating to the user can be passed between the systems. Finally, the user is given a dual-mode phone capable of working on the DECT or GSM network. The phone scans for DECT carriers and, if one is found, registers itself onto the DECT network. When there is an incoming call it passes first to the DECT system, which checks to see if the user is registered. If he or she is, then the call is terminated on the DECT network (and hence is carried over the fixed network).

If he or she is not, the call is passed to the GSM network for termination on the mobile network. Outgoing calls are, by preference, made on the DECT network.

- *Cost:* In order to provide the system, DECT base stations must be installed throughout the place of work. Although not disclosed, it is thought that the cost of this could be in excess of $1,000 per user depending on the difficulty in providing coverage in the building.

- *Call billing:* Billing for incoming calls is complicated by the fact that the calling party does not know over what network the call will terminate. As with the personal assistance, typically all incoming calls are more expensive than normal fixed line calls and, in addition, the called party may have to pay for incoming calls terminated on the mobile network.

- *Supplier capability:* This can only be provided by a PTO who either owns a mobile network or has entered into partnership with a mobile network provider.

- *Revenue potential for operator:* By charging a premium for this service, the operator can achieve increased revenue. It is also thought that users of such phones would make more calls as a result of the increased utility that they have from their phone.

- *User needs:* The user needs are broadly met by such a service. The only problem may be that the dual-mode phones are less attractive than normal mobile phones.

- *Regulatory barriers:* The only regulatory barrier is the ability of a company or the PTO to deploy cordless base stations within their offices.

Advantages   The advantage of this solution is that the user truly has a single phone on which all his or her calls arrive and from which he or she makes all calls. The user also gains mobility within the office. For the PTO, there is the advantage that mobile calls made in the office are routed over the fixed network and all incoming calls pass to them first.

Disadvantages   The key disadvantages are the cost to the corporate of installing a DECT system and of equipping the users with dual-mode phones. There is also the potential disadvantage of higher telecommunications cost as a result of having to pay toward the termination cost of some incoming calls.

6.4.3.4   Mobile substitution—the wireless office

Definition   This is similar in concept to the dual-mode phone except that the coverage in the office is provided by the mobile operator using an indoor variant of their base stations. Users can then continue to use their mobile phone wherever they are, with calls being routed to the office base stations whenever the user is in the office. Compared to the dual-mode phone, the calling party will typically experience less delay on setup and the user will have a more seamless offering that allows handover when leaving the office. The key difference is that in this scenario, the traffic that was previously carried by the fixed network is "stolen" onto the mobile network. This solution requires special in-building GSM base stations and cost-saving solutions such as backhaul compression provided, for example, as part of the Motorola Wireless Enterprise solution.

▶ *Cost:* The cost is in the provision of the in-building base station. This cost will typically be borne by the mobile operator rather than the corporate. There will be little additional cost for handsets, and the corporate will be able to remove their PBX, resulting in a saving in this area.

▶ *Call billing:* Call billing is similarly problematic as for the dual-mode solution because for incoming calls, the calling party has no knowledge of whether they will terminate in the office or on the mobile. As a result, the called party may have to pay an additional fee to terminate calls when out of the office.

▶ *Supplier capability:* This can be provided by any cellular operator and, hence, is open to more competition than almost any of the other option.

▶ *Revenue potential for operator:* This can result in a substantial revenue increase for the mobile radio operator because (1) they carry additional call traffic from within the office that should be profitable, although at a much lower level than normal mobile phone traffic; (2) they reduce churn in the corporate marketplace because most offices will only have one in-building system, locking the corporate in to the operator; (3) in the case where the corporate users were on a different operator, more valuable corporate users are added to the network who will make additional calls outside the office. For the fixed operator, the story is completely different as they will lose all, or most, revenue from this corporate. However, most fixed operators also have a mobile operation and, by transferring traffic to this mobile network, can ensure that the traffic is not lost from the group.

▶ *User needs:* This solution best meets the user needs because the user truly has one phone with a seamless and consistent service regardless of whether he or she is in the office or roaming. The phone can be more easily personalized, and there is no delay on incoming calls or problems with redirection.

▶ *Regulatory barriers:* There are no real regulatory barriers, except that only operators with a cellular license can provide this service.

Advantages   This solution is highly advantageous to the mobile operator who captures more traffic and profit. It is advantageous to the user who has a truly integrated phone service in that the user only has a single phone and a single number. Of all the FMC solutions, it probably best meets the user's needs, although it is not strictly FMC but substitution onto the mobile phone.

Disadvantages   The key disadvantages are for the fixed line operator who will lose valuable corporate traffic. Other disadvantages are the need to provide additional GSM infrastructure within the building. However, it would appear that this infrastructure soon pays for itself.

Summary   A summary of all the different solutions is provided in Table 6.2.

6.4.4   The future of the FMC marketplace

The FMC marketplace is currently exhibiting a range of solutions and opinions as to the most appropriate way ahead. This is partly due to maturing technology and partly due to the fact that both fixed and mobile operators are competing for the same (fixed) traffic. Because of this competition, at least two options will remain in place for some time, one, such

Table 6.2
Summary of the Different FMC Solutions

	Degree of FMC	Fixed Operator Advantages	Mobile Operator Advantages	User Advantages	Overall
Personal assistant	Partial, requires complex call forwarding	Slight increase in revenue, capture some mobile traffic	None	Single mailbox and increased contactability	Acceptable in the short term, but not a long term solution
Network integration	Limited, mainly billing	Reduces overall costs	Reduces overall costs if part of a PTO	Limited	Does not provide full FMC
Dual-mode phones	Full	Retain traffic that might have become mobile	None	Single phone and mobility in the office	Interesting solution but lacking support
Mobile substitution	Full	None, fixed operator is bypassed (but traffic may be carried on associated mobile operator)	Mobile operator captures fixed traffic and reduces churn	Single phone and mobility in the office	Emerging solution likely to be favored by many

as BTs OnePhone, put forward by the fixed operator, and the other, such as Motorola's Wireless Enterprise, put in place for the mobile operator.

There are also elements in all the solutions that are likely to exist well into the future. Personal assistants have value even when there is only one phone by seamlessly handling voice, data, and fax and combining this with a powerful mailbox capability. Integrated billing and operations platforms will be appropriate for operators with more than one network in the same country regardless of whether the users have more than one phone.

Looking into the future, initiatives such as the *universal mobile telecommunications service* (UMTS) aim to provide one phone for all situations. Given the benefits of mobility and the advantages of a single phone it seems likely that FMC in the future will involve some element of mobility in the office coupled with the use of dual-mode phones if necessary. Both the dual-mode phone and the wireless enterprise solutions offer the user similar levels of functionality and can probably be offered for similar levels of cost. These two solutions are likely to play the key role in FMC in the future, especially in large organizations where wireless PBXs are a practical proposition. For the SoHo and home worker, solutions might include the DECT or GSM home base stations or home billing capabilities (the Home Zone) concept provided by mobile operators.

The FMC marketplace will become a battle of two systems, both offering the user one phone. The PTO will be keen to use dual-mode cordless/cellular systems while the cellular operators will be deploying single-mode wireless enterprise solutions. There is room in the marketplace for both solutions to exist, but at present the wireless enterprise solution seems more widely supported by the manufacturers and more in tune with future telecommunications trends. Although both solutions will probably exist for some time to come, it seems likely to expect that the wireless enterprise solution will become the eventual answer to FMC, providing the user with a single phone and a seamless mobility service wherever they may be.

This brings us back to the quote at the start of this chapter. It may be that mobile networks coexist with fixed networks at the moment while it suits them to do so but will increasingly take over the functions of mobile networks in the future.

References

[]1   Clarke, M., *Networks and Telecommunications: Design and Operation*, New York: John Wiley, 1997.

[2]   Freeman, R. L., *Telecommunications System Engineering: Analogue and Digital Network Design*, New York: John Wiley, 1996.

[3]   Griffiths, J., *ISDN Explained*, New York: John Wiley, 1996.

[4]   Wilder, F., *A Guide to the TCP/IP Protocol Suite*, Norwood, MA: Artech House, 1998.

[5]   Clarke, M., *ATM Networks and Principles*, New York: John Wiley, 1996.

[6]   Rahman, M., *Guide to ATM Systems and Technology*, Norwood, MA: Artech House, 1998.

The mobile network operator

This part moves away from a consideration of the underlying technology and looks at the application of that technology to build and operate complete radio systems. Designing a network is a complicated task, requiring an understanding of cell deployment, linking of cells back into the network, and traffic theory. To get to the stage of building a network the operator needs to gain a license. The network design and operation is heavily constrained by economics, and this part describes how a business case can be put together and how the complete wireless professional can understand financial statements. Operating a network requires different skills that are also discussed in this part. Being a network operator is a skilled task, and one that should not be undertaken lightly by the complete wireless professional.

Contents

Designing a mobile radio network

Artists can color the sky red because they know that it is blue. Those of us who aren't artists must color things the way they are or people might think we're stupid.

Jules Feiffer

7.1   Technical design

7.1.1   Introduction

Most engineers working for mobile radio operators will at some point be involved in the network design process. Good design of the network is essential in order to provide a high-quality but cost-effective network. This section looks at some of the major design issues that encompass all types of mobile and fixed radio networks.

An overview of the process of setting up the network is provided in Figure 7.1.

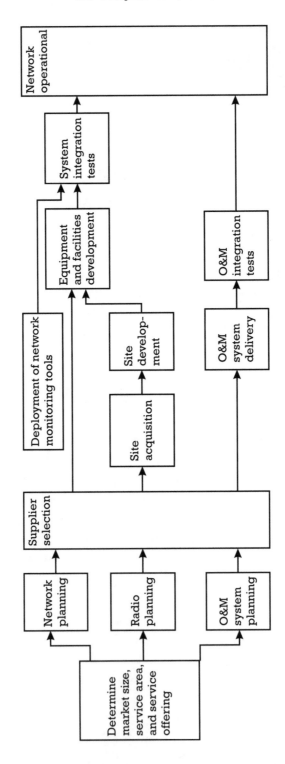

Figure 7.1   The network roll-out process.

Each of the key steps in this block diagram are now described, starting with selecting the number of cells, moving to choosing the cell sites, connecting the cells back into the network, and interfacing with switching and backbone systems.

7.1.2 Network planning

The engineer must not forget that the only purpose of building a mobile radio network is for the owner of the network to make money. A skillful network design is one that maximizes the profitability for the owner of the network. Understanding exactly how to do this requires the engineers to work in conjunction with the marketing staff over issues such as the areas for which coverage is required and the features and facilities that are to be offered. Once these have been decided, the network designer must try to design a network that achieves these requirements for the lowest possible cost.

A detailed analysis of almost any sizable mobile radio system shows that the largest element of the cost is typically related to the number of base stations, both due to the direct cost of the equipment and the indirect cost associated with the interconnection of the base stations to the switch and the site rental charges. Hence, the objective of the engineer should be to minimize the number of base stations while still meeting the network coverage and capacity requirements.

To determine the number of cells required in any given area it is necessary to assess how many cells are required for capacity and how many for coverage and then take the larger of these numbers. Put another way, there must be at least sufficient cells to provide the required coverage; if this does not provide sufficient capacity, then more cells will be needed. The manner in which the number required for coverage and for capacity are calculated is discussed in more detail, but fundamentally:

▶ The number of cells required for coverage is found from the size of the total area to be covered divided by the area of a cell given the topographical constraints in that part of the country.

▶ The number of cells required for capacity is given by the total traffic generated divided by the capacity of a cell, where the total traffic generated is the number of users in the area multiplied by the traffic generated per user.

As might be expected, this calculation will differ for different parts of the country, with city centers presenting a much higher traffic load than rural areas; hence, the calculation needs to be performed for each part of the country where these data might differ. The difficulty here is not in performing the calculation but in obtaining the input information. Topographic and demographic information is typically readily obtained from mapping sources or from a census in most developed countries. However, in developing countries there may be little idea as to the total population, resulting in the need to use estimates. Determining the expected penetration is more difficult. By examining other, similar countries, where cellular is in a more advanced state, it can be possible to make comparisons as to the expected rise in penetration. Surveys in a range of different areas can indicate the willingness of the consumers to have mobile radio systems (although note that survey design is an entire discipline in its own right, see, e.g., [1]). Traffic levels can be estimated from comparison with other countries, from looking at the fixed line traffic and fixed line tariffs, and finally from the survey process. However, it is unlikely that such an approach would have been able to predict the extremely high penetration and usage in Hong Kong, due more to social patterns rather than other indicators, and the engineer must accept that any estimates fed to him by marketing and research staff are likely to be highly approximate. It is especially worth remembering the dramatic errors made in the early days of cellular networks; for example, Vodafone in the United Kingdom predicted that their TACS network might have a total subscriber base of 100,000 at maturity. It finally peaked at over 2 million.[1]

The number of cells required for coverage in a particular area is given by

$$\#\text{cells} = \frac{\text{size of area}\left(km^2\right)}{\pi r^2} \cdot I \qquad (7.1)$$

where r is the expected cell radius in kilometers, I is a factor representing the inefficiency of tessellating cells due to the need for overlap, and the symbol # is used as shorthand for "number of."

The expected radius can be obtained from the base station manufacturer or from trial results. For deployment of systems such as GSM, the

---

1. Clearly this was in the early days of cellular, with little experience to go by; predicting cellular subscriber numbers has become much more accurate. However, for new types of networks, such as WLL, similar levels of uncertainty might exist.

cell radius in different environments is well known from the large number of deployments. For newer, or more unusual, deployments a different approach may be required. The basic calculation of cell radius starts with a link budget of the simplified form indicated in Table 7.1.

The key to the link budget is the receiver sensitivity. This is the signal level required at the receiver for adequate reception. For example, if an SNR of 9 dB is required to achieve sufficient voice quality and the noise floor is −111 dBm, then the minimum signal, or sensitivity, for good reception is −102 dBm. The sensitivity is typically supplied by manufacturers, and minimum acceptable levels can be found in the specifications.[2] There is little advantage in having an unbalanced link budget, that is, having the uplink or the downlink able to tolerate a greater path loss. When unbalanced it would be possible to reach a situation where the downlink was of adequate quality and the uplink inadequate. However, link budgets tend to be slightly unbalanced because the base station has a directional antenna unlike the mobile and because the base station can employ lower noise amplifiers that result in a lower noise threshold and increased sensitivity. In the preceding example there is a 5-dB mismatch

Table 7.1
A Simplified Link Budget Calculation*

Parameter	Downlink	Uplink
Transmitter power	+43 dBm	+30 dBm
Cable and splitter losses	−3 dB	0 dB
Antenna gain	+8 dB	+2 dB
TOTAL TRANSMIT POWER	+48 dBm	+32 dBm
Receiver sensitivity	−102 dBm	−110 dBm
Receiver antenna gain	+2 dB	+8 dB
Cable losses	0 dB	−3 dB
MINIMUM SIGNAL	−104 dBm	−115 dBm
MAXIMUM PATH LOSS	152 dB	147 dB

*The increased sensitivity on the uplink compared to the downlink is a result of better receiver units at the base station than in the mobile and is also due to the use of diversity reception at the base station.

---

2. For example, see GSM specification 05.05.

between downlink and uplink. Using the Hata model as an example, the typical propagation distance associated with this link budget can be calculated.

The Hata model is an empirically derived model of radio propagation. It predicts that the signal loss is given by

$$L(\text{dB}) = 69.55 + 26.16 \log\left[f_c(\text{MHz})\right] - 13.82 \log\left[h_b\right]$$
$$- a\left(h_m\right) + \left[44.9 - 6.55 \log\left[h_b\right]\right] \log R(\text{km}) \qquad (7.2)$$

where f_c is the carrier frequency (MHz); h_b the base station height (m); h_m the mobile height (m); R the distance from BS (km); and $a(h_m)$ the mobile antenna height correction, which differs depending on the environment. For example, for a medium-sized city

$$a\left(h_m\right) = \left[1.1 \log(f_c) - 0.7\right] h_m - \left[1.56 \log(f_c) - 0.8\right] \qquad (7.3)$$

Other mobile antenna height correction elements exist for different types of terrain. Setting $L = 147$ dB, $f_c = 900$ MHz, $h_b = 20$m, and $h_m = 2$m, the previous Hata equation can be solved in terms of R to yield a range of around 1.5 km.

The number of cells required for capacity is given by

$$\# \text{cells} = \frac{\text{traffic channels required}}{\text{traffic channels per cell}} \qquad (7.4)$$

The number of traffic channels required is given by

$$\# \text{channels} = E\left[\# \text{subs} \cdot \text{penetration}(\%) \cdot \text{busy hour Erlangs per sub}\right]$$
$$\qquad (7.5)$$

where $E[x]$ represents the conversion from Erlangs to traffic channels using the Erlang formula given in Section 2.3.

7.1.3 Radio planning

It is not sufficient simply to determine how many cells are required; appropriate cell sites must also be found. If the cell sites are not optimal, then the coverage from a particular cell may be less than expected or the overlap with a neighboring cell more than expected, resulting in the need for an increased number of cells. Given a minimum number of cells, the art of radio planning is to find the cell sites that allow this minimal number to be realized. There are two key constraints in finding cell sites, topographical constraint and financial constraints.

- ▶ *Topography:* In the case where it is desired to maximize coverage it is important to site cells on high spots since, as shown in Section 7.1.2, the antenna height has a key bearing on the range achieved. High sites might either be hills or the tops of tall buildings. However, in the cases where cells are being installed for capacity and not coverage, the converse is true and it is important to limit the range so as to minimize the interference to neighboring cells. In these cases, it will be necessary to assess how high a site is required to achieve the desired cell radius and then to search for a site meeting these criteria.

- ▶ *Financial:* In almost all cases the potential sites will be owned by individuals or corporations who will charge a rental fee for placing the base station on the land or rooftop space. In some cases, the fee charged may be unreasonable and it may be less expensive to use a different cell site that is less optimal but also less expensive.

The problem of siting cells optimally is mathematically intractable given the complexities of terrain and the difficulties with rental. Instead of attempting a theoretical solution, the most appropriate way forward is to make use of a software propagation modeling tool. Such tools take digital maps of the terrain as inputs and allow the user to place a hypothetical base station at any point. Using formulas such as the Hata formula, the model then makes a prediction of the signal strength that would be experienced. The user can then adjust the position and other parameters of the base station such as the antenna downtilt until the coverage is as required. The next base station can then be placed such that the coverage overlaps to the required degree with the first base station, again using trial and error until the best position is found. By this method, complete base station plans can be constructed surprisingly quickly.

Suitable planning tools can be purchased from a number of companies; for example, MSI in the United Kingdom makes one of the most widely used planning tools called *Planet*, while LCC in the United States is also a large vendor of planning tools. Most manufacturers also have planning tools that they can provide to users, although these are often rebadged versions of tools from companies such as MSI.

Having selected the preferred sites, a site visit is required. This determines whether it is possible to find space to site the base station and whether the site owner is agreeable to having the equipment installed. In many cases there will be problems and it will be necessary to return to the tool in order to find a different site. In some problematic cases, it may be that the lack of availability of a site forces a redesign of numerous surrounding sites; hence, it is important that no equipment is actually installed until all the required sites in an area have been finalized.

7.1.4 Microcells and picocells

The capacity of a cell is independent of its size. Therefore, small cells concentrate the same capacity into a smaller area. If a city is covered with numerous small cells rather than a few large cells the result will be a massive increase in capacity. Small city cells are referred to as microcells. The definition of a microcell is that the base station antenna is mounted below the rooftop level of the surrounding buildings. This results in the signal being constrained to the streets around the antenna. Typically a microcell will cover around 200m each side of the transmitter and up to 50m down side streets that connect onto the main street. A number of software packages are available to plan microcell networks such as NP Workplace from Multiple Access Communications in the United Kingdom or modules available within MSI's Planet tool. The prediction of microcells is relatively straightforward, the difficulty often resides in obtaining detailed digital maps of the city showing each of the buildings in the city to an accuracy of approximately 2m. Over time, such mapping data will become universally available, possibly through advances in satellite imagery, but in the short term, planning microcells can be problematic.

In some cases, cells are placed inside buildings. This is typically either to increase the network capacity by providing yet more cells or to increase the coverage inside buildings into which the signal is not propagating well from outside. In-building cells are typically known as picocells due to their very small size. The size of these cells is constrained very much by the building. Planning tools exist for picocells from Multiple Access Communications in the United Kingdom and from a number of manufacturers including Motorola in the United States. These tend to assume that the signal suffers a loss each time it passes through a wall, a floor, or a window. The tool needs to know of what each of the walls within the building are made in order to be able to apply appropriate path loss criteria. As with microcells, the availability of maps, or in this case building plans, is the key difficulty. Very few building occupants have digital plans of their buildings and even fewer know the composition of the walls and the floors.

To some extent, these problems are being overcome by the increasing availability of a cheap microcell product that does not require an optimal placement, since the penalty for using additional microcells to cover unexpected holes is not severe. Similarly, in buildings, the stage will eventually be reached where picocells will simply be placed in each corridor and in large rooms and the need for complex planning tools will be overcome.

7.1.5 Interconnection

One of the key cost elements for the base station is the connection of the base station back to the BSC and on to the switch. The cost of this connection will typically be composed of two elements: an initial cost to establish the connection and an ongoing cost to maintain the connection. The complete wireless professional must consider the sum of these two costs in determining the most appropriate means to connect the base station back to the switch. The most appropriate means to make the connection is likely to vary in different parts of the country. The key parameters that have an impact upon the choice of connection method are:

▶ The distance between the base station and the base station controller because some interconnection methods have a cost that is relatively distance independent whereas the cost of others is highly related to distance;

▶ The number of carriers deployed at the base station that will affect the capacity required from the link since some interconnect methods have a maximum capacity;

▶ The presence of available cabling because if there are not copper or fibber optic cables already in the vicinity, the cost of providing the cabling will typically be high;

▶ The availability of radio spectrum because without radio spectrum, techniques based upon radio transmission cannot be adopted.

As with the calculation of the number of base stations required, these parameters will typically vary from area to area. Often, the highest capacity will be required in city areas, which will be where fiber optic cables are readily available; whereas low capacities will be required in rural areas, where cable is not available.

There are two other factors that may impinge upon the selection of the interconnect method. The first is the desire to use the same type of interconnect as far as possible throughout the network. This is advantageous in that it reduces the need for staff training in multiple interconnect types, it reduces the need for spares holding, and it simplifies the operations and maintenance system. Deciding upon whether this is an appropriate strategy is solely a financial calculation, comparing the additional cost of using nonoptimal interconnect methods in some areas with the savings resulting from using a single interconnect method. As can be appreciated, making this calculation is often difficult, especially since it is strictly necessary to predict forward to determine the total number of "inappropriate" links over the lifetime of the network.

The second factor relates to competitive issues and is even more difficult to quantify. In some cases, especially for WLL operators, the company supplying the cabled link, typically the PTO, will be a direct competitor. The WLL company may have concerns that if they lease a service from the PTO they may experience poor service on the lines, with higher than expected outages and higher costs. There will typically be little evidence that this might be the case and hence little on which to build a financial analysis of the problem. There may also be the possibility of redress from the regulator if it is actually found that the service is not meeting the requirements. Because of these difficulties, such a decision tends to be more of a "gut feeling" than a financial decision. The complete wireless professional should try to avoid unquantifiable decisions such as this and instead seek to demonstrate the cost if the link availability proved poor and thus be able to determine the break point at which using a different interconnect methodology would be more appropriate.

The key means of interconnecting the base station sites with the base station controller use:

▶ A link leased from the PTO;

▶ A microwave point-to-point link;

▶ A point-to-multipoint overlay network;

▶ A satellite link.

Leased link   In this case, the operator obtains a link from another operator. Often the PTO, but increasingly other companies, such as railway operators, are laying cable that can be used to provide a leased-line service. The operator specifies the two end points and the data rate to be provided by the link and the rest is done by the leased-line provider. They may decide to use satellite or microwave links rather than cable to provide the link, but this will be invisible to the operator. In return, the operator pays an annual fee for use of the link. This fee is normally related to the distance over which the link runs and the capacity required from the link. There may also be an initial connection fee to link the base station to the nearest access point in the leased-link network.

In early cellular licenses in some countries, the cellular operators were required to use the PTO for all their interconnection requirements. However, this practice is now generally seen as anticompetitive and is rarely encountered.

Microwave links   These are point-to-point radio devices used to provide a connection that is similar to that provided by a cable [2]. A radio transmitter and directional antenna are required at each end of the link with a LOS path between them. Because of their LOS nature, microwave links are able to operate in higher frequency bands than mobile radio systems; however, the range does decrease as the frequency increases, as shown in Figure 7.2.

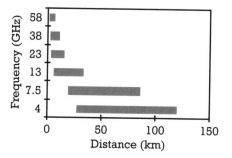

Figure 7.2   Microwave link range variation with frequency.

Broadly, a LOS path is required between the transmitter and the receiver for a microwave link to work; hence, most links tend to be composed of transmitters and receivers on tall masts or in elevated locations. Microwave links differ in their coding and modulation scheme from mobile radio systems. Typically, because of their directionality, microwave links generate little cochannel interference to other microwave links and, hence, high SNR and SIR values can be achieved. As a result, fixed links tend to use high-level modulation schemes such as 64-level QAM, linked with trellis code modulation [3]. Links are typically provided with "standard" data rates ranging from 2 to 34 to 155 Mbps.

The cost of a fixed link is the equipment and masts required. Hence, as long as the required link distance does not exceed the range of the equipment, the cost is unrelated to distance. Although higher capacity link equipment is more expensive, the difference compared to a lower capacity link is relatively small, so link cost is only slightly related to link distance. A comparison of the through life cost of a 2-Mbps fixed link with a leased line is provided in Figure 7.3. This figure will vary from country to country depending on the cost of the leased line charged by the PTO and on the availability of spectrum for fixed links.

Point-to-multipoint overlay networks   This is a relatively new concept that is undergoing trials at the moment. It is similar to the point-to-point microwave links except that, instead of having a separate dish at the base station controller for each base station, a single antenna is used to gather all the signals (in the same manner that a single antenna is used at a cellular base station to gather the signals from all the subscribers). If fact, such an overlay network is a form of WLL system, since each of the base stations looks like a fixed subscriber. Such systems are less

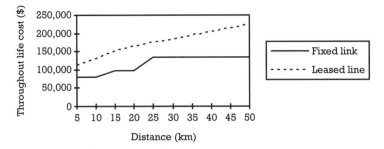

Figure 7.3   Fixed link versus leased line costs for a 2-Mbps link.

expensive than point-to-point systems since dedicated antennas and circuitry are not required for each link at the base station controller. However, they typically have a lower range both as a result of not using a directional antenna at the base station controller and because the system capacity may not be as high. This low cost but low range makes them particularly suitable for providing backhaul for microcells within a city area, and indeed some of the first applications suggested were for the backhaul of cordless base stations. The WLL system needs to provide a high capacity per base station with systems such as Ericsson's AirLine (see Section 5.3) with its 2 Mbps per user being appropriate. These are termed overlay networks because they can be considered to be radio systems sitting on top of other radio systems or overlaying them. It is likely that an increasing number of these sorts of systems will be deployed in the future, both as the technology becomes less expensive and as the number of microcells and picocells increases.

Satellite links   If the distance from the base station to the switch is greater than around 100 km and there is no existing infrastructure, then satellite systems may be the most appropriate means to provide the link. Satellite systems have the advantage that they are truly distance independent and can operate immediately in most parts of the world. Using VSATs, data rates of up to 2 Mbps can be provided. Even in the case that distance is not a problem, satellite systems are sometimes deployed. During the roll-out of the U.K. National Lottery, BT was unable to respond sufficiently rapidly to provide leased lines into each of the shops selling tickets and so around 2,000 were connected using satellite links. However, in this case, the cost was relatively low since the amount of data being transferred from each terminal was only a few tens of bits per second.

Satellite systems have the disadvantage of adding an extra delay to the signal and of a relatively high cost if significant bandwidth is required. For example, the cost of renting a 2-Mbps leased line is approximately $500,000 per year, which is normally much more expensive than the other options described previously.

7.1.6   Operations and maintenance planning

The *operations and maintenance* (O&M) system typically requires little in the way of planning. It will be necessary to understand the size of the network in terms of number of cells in order to be able to order an O&M with adequate capacity. Another important point will be to determine how many network control rooms are required, whether they will be located close together, and whether some will be shut down during off-peak periods in order to reduce staffing numbers. It might also be necessary to consider future network upgrade plans, such as a plan to add packet data transmission at a later date. Clearly, it will need to be possible to reconfigure the O&M system in order to add the capabilities.

7.1.7 Supplier selection

Supplier selection is a time-consuming and complex area that covers both technical and financial matters. A textbook supplier selection process would cover the following stages.

1. Detailed understanding of the requirements of the equipment based upon preliminary network design, prediction of future requirements, and an understanding of all the services that are to be offered.

2. Development of a *request for quotations* (RFQ), which is a formal document setting out the requirements on the equipment. Each requirement is listed, along with an indication of whether it is mandatory or optional. The writing of such a document is highly time consuming because it must cover everything from lightning protection and earthing requirements, to equipment size, to all aspects of equipment function.

3. The RFQ is sent to a short list of preferred suppliers.

4. The quotations received from the suppliers are formally evaluated. Each quotation may be scored against the number of requirements that it is able to meet. Then there follows the difficult process of comparing the prices offered against the functionality "score." Often the highest functionality will have the highest price associated with it.

5. Contract negotiations commence with the preferred supplier.

6. Once the contract has been signed, the equipment manufacture and delivery can commence.

It is often tempting to shorten these procedures, especially when procuring commodity items such as GSM base stations. Less formal systems can save substantial amounts of time but can lead to problems at a later stage. Many operators make use of consultants during the supplier selection process in order to provide the necessary man-power and expertise. Selecting a consultant who already has a template RFQ can save substantial time and effort.

7.1.8 Network deployment

The deployment of a cellular network, once the sites have been selected and the equipment has been procured, is essentially a large-scale project management task. The operator might be rolling out one site a day in order to meet coverage requirements. Each site requires a survey, civil works to erect the mast, the provision of power and communications, the delivery of the base station to the site, the erection of the mast, the installation of the antennas, and the commissioning of the base station. This will require a wide range of contractors from civil engineers to antenna riggers. Ensuring that each person turns up to the site when required is a major logistics exercise. Increasingly cellular operators have been turning to large civil engineering companies to provide the project management skills required for such a venture. Such companies will have project management experts, a range of project planning tools, and the required experience. The difficulties of network deployment should not be underestimated, but this is a task that the complete wireless professional can delegate to those with more appropriate skills, only becoming involved during commissioning and test and when unexpected problems arise.

7.2 Applying for a license

A license application is generally required before an operator is provided with radio spectrum and a license to run a telecommunication service. General factors to include within the license application are:

▶ Introduction;

▶ Overview of the operator including ownership and turnover;

▶ Details of the service proposed;

- A technical description of how the network will be realized;

- A marketing plan showing the services that will be offered and any new and innovative means for provision of the mobile radio service (e.g., the use of prepayment mechanisms);

- A demonstration that the radio spectrum required is being used efficiently;

- Simplified business plans showing the justification for building the network;

- A network build plan showing the speed with which the network will be built;

- A summary of the advantages to the country;

- Appendixes including coverage maps, detailed technical information, details of the frequencies required, and interference assessments with other users of the radio spectrum where required.

The most important fact to understand is what the regulator or government wishes to get from the license award. Possible government objectives will include:

- Increasing the competition for telephony in the country;

- Increasing the wealth of the country through enhanced provision of communications;

- Providing telephony to rural regions;

- Stimulating local industry and improving local skills;

- Gaining a cash injection through the sale of radio spectrum;

- Providing higher bandwidth services, allowing further economic development.

It is a little realized fact that the best license applications are often produced by small operators with apparently limited resources while the worst license applications are often produced by the large multinational partnerships who, it would seem, should have both more experience and greater resource than the smaller companies. There are many reasons for this, but key is the quality and number of staff responsible for writing the bid. In the same manner that books written by one or two authors typically have more continuity than edited books where each author has written a different chapter, license applications written by small closely knit teams are typically better than those written by very large teams from a range of different organizations with different cultures and values. The large companies also are often rather arrogant about their power and perhaps do not assign the quality of staff required to the process. The best way to write a license application seems to be to assemble a small team of half a dozen highly skilled individuals, experts in their own fields and with good writing skills, and to provide them with whatever support they desire (such as technical staff to produce some coverage plans).

A good license application should read like a good book, with clear logical flow and lack of repetition. It should avoid providing information well-known to the assessors, like the architecture of the GSM system, and concentrate on describing the novel and advanced features that will be offered in the network. A common trait in license applications is to start sections with "Company x will provide a network of the highest quality with exceptional coverage and capacity," but then not to substantiate this statement in any way. A good license application would then go on to explain how microcells will be widely deployed, showing that the cost and coverage area of a microcell was fully understood, then developing some traffic models, and showing how the network capacity would be well in excess of these traffic models.

It is becoming increasingly true that licenses are won not on technical capability but on the new services that will be offered by the network and on the new means of marketing and providing good customer support. Most assessors now assume that any large organization can muster the technical skill to assemble what is becoming a commodity—a GSM network. Licenses have been won based on a commitment in the license application to reimburse customers for any calls that were dropped—something that requires little technical skill but an understanding of the cost of such a commitment. The importance of engineers in writing the license application is therefore diminishing for all but the new and unusual networks (such as WLL deployments or dual-mode systems). Engineers should realize this and ensure that they have all the marketing and sales support that is required.

7.3   The mobile radio
equipment manufacturer

Many engineers will work for manufacturers rather than operators in a range of roles including equipment design and manufacturers, research and development, antenna design, test equipment design, software tools for mobile radio system, and RF component design, for example. The basic role of the manufacturer is to provide the equipment required by the operator for them to provide a profitable service. However, this is far from simple as the operators often do not know exactly what they want nor what is possible. Design cycles are often four years or more long; hence, the manufacturer will need to develop products well before they are required by the operators. This is evidenced in standards forum where the vast majority of contributors are the manufacturers. The only operators to attend are typically those that are also PTOs and have a large research facility.

In a world that is increasingly dominated by standards, the manufacturer will seek to work in the following manner. First, they will conduct research around the areas on which standardization is required and will seek patents on any key breakthroughs. They will then try to drive the standard in the direction of their patent so that they hold as much of the key IPR as possible around the standard. This ensures that they will not have problems with other manufacturers charging high levels of payment to use their patents and raises the potential that the first manufacturer will themselves be able to benefit from royalty payments. For these reasons, the standards forums can become quite heated, as explained in more detail in Chapter 13.

Once the standard has been completed, the manufacturer will work as quickly as possible to bring the product to market. They will also be looking for ways to differentiate their product from the competitors while still conforming to the standard. In the area of handsets, this might be by making them smaller, with longer battery life than the competitors. In the case of the network, this might involve lower power consumption, greater sensitivity, and requirements for less communication resources, for example. In the world of GSM infrastructure there are hundreds of ways in which products from different manufacturers are differentiated.

Manufacturers will also seek to prime markets. A good example of this is in the area of in-building coverage. This is an area that the manufacturers would very much like to develop since it would allow them to sell increased volumes of infrastructure for installation in office buildings around the world. The manufacturers cannot sell this product until the operators are convinced of its benefits. Hence, the manufacturers need to perform the business case and the trials in order to demonstrate the viability to the operators. Once the market has been "seeded," the manufacturer can take a less dominant role since the "pull" from the operator will become sufficient to develop and sell product.

Some manufacturers are increasingly taking on a role as operators. In most cases, this is somewhat unwillingly because historically manufacturers have tended not to be successful as operators due to the different skills required. However, today's operators are expecting vendor financing (see Section 8.2.5) and for the manufacturer to *build, operate, and transfer* (BOT) the network to them. The result is the manufacturer performing the network design and initial network operation and maintaining a position on the board of the operator so that they can see that their investment is being safeguarded. It is not clear where this trend will lead in the future, and indeed it may be unhealthy for competitive supply if manufacturers take too large a stake in operators.

Manufacturers are starting to see the cellular marketplace changing. Until recently, cellular remained a complicated and knowledge-intensive industry where only those manufacturers with background expertise in communications and an excellent research and development capability could flourish. These manufacturers were able to charge a premium price as a result of their expertise. Now, with the increasingly widespread availability of chipsets for most of the major wireless standards, it is comparatively simple to become an equipment provider and many electronics companies are entering this field. Just like radios and televisions, cellular is now becoming a commodity product with the effect that manufacturers are continually reducing prices. The manufacturers are trying to differentiate themselves by having better specifications or wider features lists and by opening new markets such as WLL and in-building where their superior technical expertise will allow them to charge higher prices in the short to medium term.

References

[1]   Crimp, M., *The Market Research Process*, New York: Prentice-Hall, 1990.

[2]   Freeman, R., *Radio System Design for Telecommunications (1-100GHz)*, New York: John Wiley, 1987.

[3]   Webb, W., and L. Hanzo, *Modern Quadrature Amplitude Modulation*, New York: John Wiley, 1994.

CHAPTER 8

Economics of a mobile radio network

Money is what you'd get on beautifully without, if only other people weren't so crazy about it.

Margaret Case Harriman

8.1 Understanding financial information

8.1.1 Introduction to accounting

Engineers typically do not find accounting interesting. Compared to the ability to design a network and see something realized and working, the compilation of numbers in a strict and unbending format has little allure. This is not unduly problematic as engineers will rarely be asked to write the accounts. However, engineers are often asked to look at the accounts and to contribute information to the business planning process. When the business plan does not show an adequate return on investment, the engineer may be expected to find savings in the network design. An understanding of how to read and analyze conventional ways for representing monetary flows is an important attribute for the complete wireless professional. There are many books that can be used to provide a guide to the manner in which accounts are compiled; however, the wireless professional has no need for those that tell the reader how to actually perform the accounting process. Some useful reference works for those looking to delve deeper into the accounting process include [1–3]. In this section, a simple guide is provided to the three main parts of the accounts: the profit and loss statement, the balance sheet, and the funds flow statement.

It is important for the complete wireless professional to understand the standard accounting format used around the world and the means whereby a business plan is put together.

8.1.2 The profit and loss account

The prediction of the profit and loss account for a network lists the incomes and expenditures for a business for each year of operation. The company accounts will show the profit and loss account for the year just past. It shows how much profit has been made and how that profit is used—whether paid out to the shareholders or retained for future use. The profit and loss account basically adds up all the income and then subtracts from that all the costs. A fabricated (but nevertheless approximately accurate) profit and loss account for a cellular operator projected for the first eight years of operation is shown in Table 8.1.

Most of this is relatively self-explanatory. Income is made up of connection and subscription fees and call usage fees. It also arises from call fees paid by roaming subscribers and interconnect revenue (explained in more detail in Section 8.2.4) that is paid to the cellular operator when an incoming call is terminated on a mobile phone. Against this is balanced the costs of providing this service mainly related to the handset subsidy and the cost of acquiring subscribers. There is also an interconnect cost associated with calls that are terminated on other networks. Note at this point that nothing has been said about the cost of providing the network itself. The difference between the income and the costs directly associated with that income is called the net revenue.

Table 8.1
A Typical Profit and Loss Account

	1998	1999	2000	2001	2002	2003	2004	2005
Revenue	All in US$ million/year							
Connections	0.3	0.9	0.8	1.0	0.9	0.6	0.4	0.2
Subscriptions	2.2	6.9	10.8	16.0	20.6	24.0	26.9	29.0
Call usage	3.4	12.7	22.4	30.6	38.1	43.4	46.5	47.3
Roaming fees	0.8	1.0	1.3	1.6	1.9	2.1	2.4	2.5
GSM services	7	22	35	49	62	70	76	79
Interconnect revenue	0.7	2.7	4.9	6.9	8.9	10.5	11.9	12.5
Total gross revenue	**7**	**24**	**40**	**56**	**70**	**81**	**88**	**92**
Direct costs								
Handset subsidy	1.0	2.5	2.2	2.4	1.7	0.9	0.2	0.2
Sales costs	0.2	0.7	1.2	1.7	2.4	4.7	5.1	5.3
Interconnect costs	0.1	0.5	0.9	1.4	2.0	2.5	3.0	3.5
Total direct costs	1	4	4	6	6	8	8	9
Net revenue	**6**	**21**	**36**	**51**	**64**	**73**	**80**	**83**
Operating expenses								
Maintenance	1.7	1.9	2.2	2.5	2.8	3.1	3.3	3.4
Leased lines	0.0	0.0	0.0	0.1	0.1	0.1	0.1	0.1
Sales and marketing	1.2	3.3	3.0	3.5	3.6	3.1	3.2	2.7
Staff	1.6	4.1	2.5	2.9	3.0	3.4	3.6	13.2
Accommodation	0.2	0.3	0.4	0.5	0.6	0.8	0.9	1.6
General and administrative	0.9	1.2	1.7	2.2	2.8	3.3	3.7	4.2
Consultancy	0.6	—	—	—	—	—	—	—
Regulatory fees	3.9	1.8	1.8	1.8	1.8	1.8	1.9	1.9
Bad debts	0.0	0.1	0.2	0.3	0.4	0.4	0.5	0.5
Contingency	0.3	0.4	0.4	0.4	0.5	0.5	0.5	0.8
Total operating expenses	**10**	**13**	**12**	**14**	**16**	**16**	**18**	**29**
Operating profit	(4)	7	24	36	49	56	62	54
Less: Depreciation	4	10	11	13	14	15	16	17
Net profit before interest and tax	**(9)**	**(2)**	**12**	**24**	**34**	**41**	**46**	**37**

Table 8.1 (continued)

	1998	1999	2000	2001	2002	2003	2004	2005
	All in US$ million/year							
Interest payments (receipts)		3	3	3	3	2		(1)
Tax			5	8	11	14	16	14
Net profit after interest and tax	(9)	(5)	4	12	20	26	30	24
Less dividends					5	17	14	15
Retained earnings	(9)	(5)	4	12	15	8	16	9

The indirect or operating expenses such as the costs of leased lines and office space are offset next. The net revenue less the total operating expenses forms the first of a number of definitions of profit (so if someone just quotes a profit, be clear to which part of the profit and loss account it refers). This profit is called the operating profit and determines whether a profit is made when the income is compared to the direct and indirect costs of providing that income.

The next step is to subtract depreciation. Depreciation is a slightly confusing accounting convention and is worth explaining in more detail. Imagine a generally profitable company that decides one year to replace its main building. If the cost of the building was placed directly on the profit and loss statement, the company would show a loss for that year, despite the fact that the business was profitable and the building would contribute to profits over the coming years. Investors might lose faith in the company unless it was explained to them that the company was still profitable but just had a large piece of expenditure. The argument would then be that if the company is profitable why does the profit and loss account not show a profit? The convention of depreciation is a way around this problem. Instead of placing the whole of the cost of the building in the year it was built, the cost is spread across the expected lifetime of the building, normally in a linear fashion (exponential or quadratic depreciation is sometimes used, but normally judged to be too complex). Because judgments about the lifetime of certain items, coupled with possible resale value, become complex and open to "creative accounting," there are a number of rules that determine how depreciation should be approached.

There are plenty of misconceptions surrounding depreciation. The fact that an object is depreciated does not mean that any money has been saved for a replacement. If an object has been fully depreciated and is still in use, it does not mean that it should be disposed of, just that the original estimate of its lifetime was incorrect. In fact, a fully depreciated item is no different than a partly depreciated one, except in the manner it is represented in the profit and loss statement. Sometimes depreciated items are disposed of before they have lost all their usefulness purely for tax purposes so that a new item can be brought, the depreciation on the profit and loss account increased, the profit decreased, and hence the tax liability reduced. This need not concern the complete wireless professional, only the accountant.

As can be imagined, depreciation is an important issue for the mobile radio operator who has to buy and install a high-cost network in the early years of operation. Typically, the cost of this network is depreciated over 10 or 15 years, resulting in a relatively small entry in the profit and loss account in the early years (10% of the actual expenditure during the year in this example). Because of depreciation, the profit and loss account does not reflect whether there is money in the bank (because a profit after depreciation would not necessarily be a profit after the full expenditure is taken into account) and, hence, the profit and loss account cannot be used to estimate the funding requirement.

After the depreciation has been subtracted from the operating profit, what remains is the net profit before interest and tax. This is sometimes abbreviated *earnings before interest and taxation* (EBIT). Subtraction of the interest and the tax payable results in the profit after interest and taxation. The company then makes a decision as to what it is going to do with this profit (or loss), typically either pay it to the shareholders or retain it for future purposes.

8.1.3 The balance sheet

The balance sheet is of less interest to the wireless professional than the profit and loss account. The balance sheet shows what the company owns and how it is financed. It is useful in determining the value of the company and checking on how much money it owes. It is of no use in determining whether the operation is profitable and little use in determining whether a business plan is appropriate—the profit and loss account is best suited to this purpose. A balance sheet for the same cellular operator is shown in Table 8.2.

The balance sheet seeks to assess everything the company owns. Clearly it owns infrastructure such as base stations and towers, which are known as fixed assets (because they are fixed in space and difficult to sell quickly). Each year more fixed assets are added but against this, the existing fixed assets lose value as they become older and this loss of value is calculated by subtracting the depreciation from the total value of the network. This results in the sum of fixed assets less accumulated depreciation.

Below this comes current assets. These are short-term resources that are either cash or expected to become cash in the next year and they typically comprise stocks (work in progress), debtors, and cash. In the case of a cellular operator there is little in the way of stock. Debtors represent unpaid bills at the end of the year (typically the bills from the last month of operation before the date of the balance sheet), and then cash is split between cash in the bank and cash held in short-term deposits (where it can earn a higher rate of interest).

It is then necessary to subtract from this the liabilities: money or goods owed by the company. In this case, the operator owes its creditors money

Table 8.2
A Typical Balance Sheet

	1998	1999	2000	2001	2002	2003	2004	2005
Fixed assets	All in US$ million/year							
Fixed assets at year start	—	85	99	113	127	144	155	166
Additions: Network	84.2	13.2	14.5	14.0	16.1	10.5	10.9	8.2
Towers	1.0	0.1	0.1	0.1	0.1	0.1	0.1	0.1
Office equipment	0.2	0.0	0.0	0.0	0.0	0.2	0.0	0.0
Total additions	85	13	15	14	16	11	11	8
Total fixed assets at cost	85	99	113	127	144	155	166	174
Less: Cumulated depreciation	4	14	25	38	52	68	84	101
Fixed assets less accumulated depreciation	81	85	88	89	91	87	81	72

Table 8.2 (continued)

	1998	1999	2000	2001	2002	2003	2004	2005
Current assets	All in US$ million/year							
Stocks	—	—	—	—	—	—	—	—
Debtors	0.4	1.6	2.8	3.8	4.8	5.4	5.8	5.9
Cash at bank	1.7	2.2	2.0	2.4	2.6	2.7	2.9	4.8
Short-term deposits	—	—	—	—	—	—	18.0	36.0
Total current assets	2	4	5	6	7	8	27	47
Current liabilities								
Creditors	12	3	3	3	3	3	3	3
Dividends payable	—	—	—	—	5	17	14	15
Total current liabilities	12	3	3	3	9	20	17	18
NET CURRENT ASSETS	(9)	1	2	3	(1)	(12)	10	28
TOTAL ASSETS	72	86	90	93	90	75	91	101
Financed by:								
Bank loans	40	50	50	41	23	—	—	—
Owner's equity								
Shareholders capital	40	50	50	50	50	50	50	50
Reserves	—	—	0	1	2	3	5	6
Retained earnings	—	−9	−14	−10	1	15	22	37
Profit (loss) current period	−9	−5	4	11	14	7	15	8
Total owner's equity	**32**	**36**	**40**	**52**	**67**	**75**	**91**	**101**
TOTAL CAPITAL AND RESERVES	72	86	90	93	90	75	91	101

(typically delayed payments to providers of equipment) and payments promised to shareholders which have not yet been paid. The total of current assets minus current liabilities forms the net current assets and, when this is added to the fixed assets, provides the total assets of the company. This is one-half of the balance sheet.

The second half shows where the money has come from to pay for these assets. This is divided into loans (bank loans) and then the money that the owners have put into the business, mainly from payments by shareholders purchasing shares. In addition, there is money transferred to reserves, earnings retained from previous years, and the profit from the current year that can all be used to finance the costs of the network. The net current assets should equal the total capital and reserves, hence the balance sheet is said to balance. If it does not, then there are either some assets that apparently do not belong to anyone or some additional financing that has apparently been lost.

8.1.4 The funds flow statement

The funds flow statement is rendered necessary by using depreciation in the profit and loss account. As a result, the profit and loss account does not show how much money is flowing in or out of the bank at any time. The funds flow statement does just this—it shows where money is flowing in and out and hence allows the maximum funding requirements to be determined so that suitable financing can be agreed with banks or whatever. The engineer will have very little concern with the funds flow unless there is a maximum funding requirement that needs to be met. An example of a funds flow statement for the same mobile radio operator is shown in Table 8.3.

The way that this is calculated is moderately interesting. The cash inflows are profit (from the profit and loss account) plus depreciation (since this is never "paid" in any case) plus any interest. The outflows are the capital expenditure, interest payable, any change in working capital (money held in the bank to pay items such as salaries), and tax payable. The difference between these two shows whether additional cash is required on an annual basis. As can be seen, after the third year, the business starts to generate more cash than it consumes (becomes "cash positive"). Clearly, the requirement for cash needs to be met. In this case, the funding comes from the shareholders and from the bank; this part of the funds flow shows the additional funding required each year, with half coming from the shareholders and half from the bank.

Table 8.3
A Typical Funds Flow Statement

		1998	1999	2000	2001	2002	2003	2004	2005
Cash inflows		All in US$ million/year							
	Profit before interest and tax	−9	−2	12	24	34	41	46	37
	Plus depreciation	4	10	11	13	14	15	16	17
	Interest receipts	—	—	—	—	—	—	—	1
	Total cash inflows	−4	7	24	36	49	56	62	55
Cash outflows									
	Capital expenditure	85	13	15	14	16	11	11	8
	Interest payments	—	3	3	3	3	2	—	—
	Working capital increase/ (decrease)	−9	10	1	1	1	1	0	1
	Corporation tax	—	—	5	8	11	14	16	14
	Total cash outflows	76	27	24	27	31	27	27	23
CASH SURPLUS/ (FINANCING REQUIRED)		−80	−19	−1	9	17	29	35	32
Funded by:									
	Shareholders equity	40	10	0	—	—	—	—	—
	Shareholders advances	—	—	—	—	—	—	—	—
	Bank loans	40	10	0	—	—	—	—	—
	Short-term deposits	—	—	—	—	—	—	—	—
	Total annual funding	80	19	1	—	—	—	—	—

Table 8.3 (continued)

	1998	1999	2000	2001	2002	2003	2004	2005
All in US$ million/year								
Used for:								
Repayment of bank loans	—	—	—	9	17	23	—	—
Surplus distributed as:								
Dividends	—	—	—	—	—	5	17	14
Transfer to short-term deposits	—	—	—	—	—	—	18	18
Total	—	—	—	9	17	29	35	32
CUMULATIVE CASH SURPLUS/FINANCING REQUIRED	−80	−99	−100	−91	−73	−45	−9	22
Cumulative funding								
Shareholder funding	40	50	50	50	50	50	50	50
Shareholders advances	—	—	—	—	—	—	—	—
Bank loans	40	50	50	41	23	—	—	—
Total cumulative funding	80	99	100	91	73	50	50	50
CUMULATIVE CASH RESERVES	—	—	—	—	—	—	18	36

Once there is a cash surplus, it is important to understand for what it is used. In this case it is used first to pay back the bank loans and is then paid as dividends to shareholders, with some additional cash used for short-term deposits. Only by the eighth year are all the debts paid and the company actually has cash in the bank. The final section shows the cumulative bank loans required, where it can be seen that loans of up to $100M in the third year will be required, with all bank loans being paid back by the sixth year. Clearly, the bank will be very keen to examine the funds flow statement.

8.1.5 Performing first pass modeling

Although wireless professionals are rarely asked to build complete financial statements, they may be asked to perform some first pass modeling in order to assess the practicality of certain situations. This section describes how this is done using a real-life example (although all of the specific details have been changed to preserve commercial confidentiality). The key is both in knowing how to provide the resulting information and in understanding when assumptions can be made and when sensitivity analysis is required. The example is based on the introduction of a railway GSM service into a public GSM network; for more details on the railway GSM service, see Section 10.2.

At the highest level, this sort of analysis compares the inflows with the outflows over a period of time and determines whether the operation will be profitable. In the case of this example, the outflows are relatively clear; the network must be enhanced to provide the services needed by the railways. Discussion with the railways showed that the following changes to the network would be required.

▶ Coverage would be required along some railway lines that would not previously have been covered.

▶ The additional features of group and broadcast call and packet data provision would need to be added to all base stations covering the railway lines.

▶ Base stations covering the railway lines would need to be made more reliable through the provision of a standby transmitter that could be switched on should the main transmitter fail.

Determining the cost of all these additions was not straightforward. In determining the cost of additional coverage, the number of additional base stations needed should have been calculated. However, in a first pass assessment of this sort there typically is not the available resource to perform cell planning along the railways. Instead, an estimate was made as to the coverage radius of a typical railway cell and then the length of track to be covered was be divided by the cell size (twice the radius) to arrive at an estimate of the number of cells. When attempting to cost the additional features there was the difficulty that manufacturers were still developing the features and, therefore, unable to provide an estimate of their additional cost. The best approximation was to look at the additional cost of features that seemed to be of a similar complexity and had already been provided by the manufacturers. Fortunately, the cost of a standby transmitter could be accurately determined.

Next, the timescales over which these new features are to be deployed were determined, and as a first pass the cost of the new features was spread linearly across this timespan. When calculating the cost it was necessary to include the cost of maintenance in terms of engineering staff and spare parts, the cost of transmission from the base stations, and the cost of site rental and power. The last three form ongoing costs that would last for the lifetime of the project.

Having determined the costs, the revenues could be calculated. However, at this stage, the price to be charged to the railways was not known and would remain a subject for negotiation for some time. The first pass analysis looked at the feasibility of the project by asking "if the railways continued to pay the same levels as they currently do for their existing network, would the GSM network be profitable?" Existing costs could sometimes be obtained from the railways; if not, estimates were made based on the number of base stations and the equipment deployed. The current payment was then compared with the GSM costs to determine whether it was worth proceeding further with the negotiations and, if so, what the lowest acceptable payment from the railways would be.

When evaluating the overall monetary flows it is important to look at the NPV of the project (NPV is discussed in Section 8.2.6). This allows appropriate compensation for the cost of capital and the likelihood that the operator will need to invest money in the first place before that cost can be recouped through annual payments from the railways. Having performed this analysis, it is instructive to examine how sensitive the results are to the assumptions made. For example, the cost of the enhancements and the number of base stations required were both estimated. By changing these figures by, say, ±10 and ±20%, the impact on the payments required from the railways can be seen. If there is a significant sensitivity to this variable, then it would be appropriate to investigate the likely value of the variable in more detail by, for example, performing some modeling of coverage along the additional railway lines.

8.2 The business case

8.2.1 The overall structure of the business case

The business case is typically a spreadsheet model that shows how the money to be invested in building a telecommunications network will be repaid. The business case reveals whether an attractive rate of return can be earned on the capital and, as a result, whether the project should be undertaken. The business case is effectively the profit and loss projection as described in Section 8.1; here, we look at the way that the constituent parts of that projection are developed.

Broadly, the business case subtracts total expenditure in the form of capital and operating expenses from the total revenue derived from the subscribers. The difficulty lies not in performing the mathematics but in deriving accurate estimates of expenditure and revenue. In particular, it is in the area of expenditure where the greatest difficulty lies. Estimating subscriber revenue is not something covered in detail in this book—it belongs more to the area of forecasting and market research—but fundamentally, predictions must be made about subscriber numbers based upon historical growth for other operators, penetration levels, analysts' predictions as to future subscriber numbers, and an understanding of demographic, social, and technical trends.

8.2.2 The network costs

One of the key cost components is the capital cost associated with the purchase of the network. Similar to network equipment, most cellular operators now purchase the mobile phones and subsidize their sale to the public. Hence, it is necessary to consider the sum of both the infrastructure and the subscriber unit costs.

Subscriber unit costs   The basic cost equation here is to multiply the difference between the manufacturer's price and the subsidized price by the volume of units sold. This can be complicated by the fact that there is typically a wide range of mobiles, each mobile attracting a different amount of subsidy. The key elements of the equation are:

▶ *The number of mobiles subsidized each year:* This will equal the predicted gross number of new subscribers each year and the extent to which mobiles are subsidized. (*Gross subscriber numbers* are the total number of new subscribers joining the network in the year. *Net subscriber numbers* are equal to the gross numbers minus the number of subscribers leaving the network.) In some countries no subsidy is paid and the user must pay for the handset in full. In other countries a full subsidy is provided and the user gets the handset for free. It is in the subsidy that the impact of "churn" (i.e., users leaving one network to join a different network) is felt.

▶ *The projected cost of the units over the investment period:* In most cases, mobile costs fall as manufacturers move along the learning curve. Estimating future falls is difficult and manufacturers will be unwilling to commit to aggressive cost reductions. Historical trends can be useful and they typically show steep price falls when a new technology is introduced, with price levels falling increasingly slowly as the technology becomes mature.

▶ *The sales cost:* This cost is composed of the advertising and distribution costs of signing new subscribers. This cost includes the costs of the shops used to sell subscriptions and the time of the sales staff in these shops.

A typical spreadsheet for the calculation of the mobile subsidy cost might look like Table 8.4.

The length of time over which the investments will be considered is not always obvious. Typically, large infrastructure projects (which a telecommunications network is) are considered over the lifetime of the infrastructure. Although a telecommunications network will probably last for around 15 years, advances in technology might make it redundant long before that. As a result, some investors only look over a 5-year horizon. However, as mobile communications becomes more competitive, returns in less than 5 years are increasingly rare and a compromise of a 10-year investment window is often used. A 5-year cycle has been shown here to allow the tables to fit easily onto the page and not to complicate the analysis.

Table 8.4
An Example of a Spreadsheet for Subscriber Unit Calculations

Basic assumptions	(all prices in $)	
Consumer mobiles	300	
Business mobiles	500	
Cost of sales	100	
Annual price reduction		7%
Subsidy		50%

	Year					
	1998	1999	2000	2001	2002	2003
Mobiles sold/year (in thousands)						
Consumer mobiles	50	150	300	500	500	300
Business mobiles	15	45	90	150	150	90
Consumer cost ($)	300	279	259	241	224	209
Business cost ($)	500	465	432	402	374	348
Annual cost ($million)	29.0	82.3	155.8	246.0	233.3	132.9

Network costs The network costs are comprised of:

1. Base stations;

2. Base station interconnection;

3. Base station controllers (if required);

4. Base station controller interconnection (if required);

5. Switching costs;

6. Operations, maintenance, and billing system costs.

For each of the capital items (i.e., items 1, 3, 5, and 6) there will be hardware costs, installation costs, and planning costs.

▶ The hardware cost will be provided by the manufacturer. It is often important to take account of all the cards needed to fully populate the equipment racks when determining hardware price.

▶ The base station and base station controller interconnection (i.e., connecting the base station to the switch) cost is dependent on the interconnect method used (see Section 7.1.5). Leased lines result in little capital charges but substantial operating expenditure. Microwave links have the opposite effect on the business case.

▶ The switching cost elements are typically comprised of the hardware cost of the switch, the accommodation costs, and the costs of providing back-up power supplies and security. Typically, the failure of a switch impacts heavily on the revenue since a large part of the network will be out of operation and a range of measures are taken to provide redundancy and ensure a very low level of failure. The O&M systems will typically be costed with the switch.

Table 8.5 shows an example of a spreadsheet for a simple system where base station interconnection is not required.

Table 8.5
Example of a Spreadsheet for Network Cost

Basic assumptions	(prices in $'000)
Base station cost	50
Installation	20
Planning	5
Base station interconnect via microwave links	
Cost	70
Installation	10
Switch costs	1,000
Switch installation	1,000

	Year				
	1998	1999	2000	2001	2002
BSs installed	59	177	354	591	591
Switches installed	1	1	2	3	3
BS total cost ($m)	9	27	54	91	91
Switch total cost ($m)	2	2	4	6	6
Total cost ($m)	11	29	58	97	97

8.2.3 The operating expenses

As well as capital expenditure, there will be ongoing operational expenditure in the form of:

▶ Rental for the base station and base station controller sites;

▶ Leased-line costs where used;

▶ Maintenance on all the network equipment;

▶ Costs of radio spectrum;

▶ Subscriber management costs;

▶ General management costs including buildings and facilities;

▶ Marketing, sales, and subscriber retention costs.

Most of these are either self-explanatory or have been covered elsewhere. Worth noting is the cost of maintenance, which is typically around 1% of the cumulative capital expenditure; the cost of radio spectrum (see Section 12.3), which in some cases can exceed all other expenditure; the subscriber management costs, which include call centers and sending bills to the subscriber; and the subscriber retention costs, which may include discounts for long-term users or periodic free replacement phones.

An example of a spreadsheet showing how some of these costs might be modeled is provided in Table 8.6.

In this highly approximate example, where the base stations have been interconnected by leased lines, the key costs are maintenance and leased lines.

8.2.4 Revenue

This is where any attempt at science disappears and art takes over! In principle, the calculation of revenue is methodological and is shown in overview in Figure 8.1.

The various parts of this figure are described throughout this section. In essence, the predicted number of subscribers is multiplied by the call minutes per subscriber to derive the total call minutes. These need to be segregated according to whether they were made in peak or off-peak periods. They need to be divided into incoming and outgoing calls. Finally,

Table 8.6
Example of a Spreadsheet Showing Ongoing Expenses

Basic assumptions		
Average site rental	($'000/year)	5
Cost per leased lines	($'000/year)	15
Maintenance	(%)	2.5
Radio spectrum	($'000/year)	1,000
General costs	$/sub/year	10
Marketing costs	$/sub/year	10

	Year				
	1998	1999	2000	2001	2002
Background information					
Total subs ('000)	65	260	650	1,300	1,950
Total sites	59	236	590	1,181	1,772
Ttl network cost ($m)	11	40	99	197	294
Ttl sub unit cost ($m)	29	111	267	513	746
Ongoing costs ($m)					
Rental	0.2	1	2	5	8
Leased lines	0.8	3	8	17	26
Maintenance	1	3	9	17	26
Spectrum	1	1	1	1	1
General costs	0.6	2	6	13	19
Marketing costs	0.6	2	6	13	19
Total ($m)	4	14	35	68	101

they need to be divided according to their destination, with possible destinations to:

▶ The national PSTN;

▶ An international PSTN;

▶ Another mobile on the same network;

▶ Another mobile on a different network.

Each of these different classes of call will attract a different tariff and a different interconnect charge. Hence, for each of the different classes of

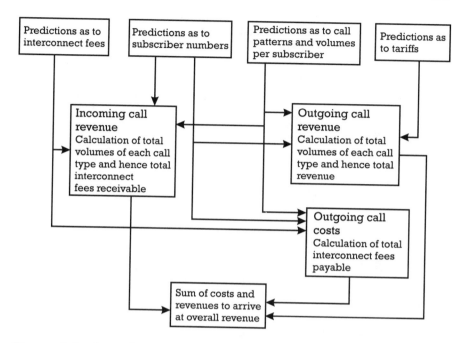

Figure 8.1 Overview of the calculation of revenue.

calls the total call minutes is multiplied by the tariff and the interconnect costs as shown in Figure 8.1. The total revenue is equal to the total call fees charged to the subscriber less the interconnect costs for those outgoing calls terminating on a different network but plus the interconnect costs for those calls originating outside the mobile network and terminating within.

This much seems reasonably scientific and can readily be encoded into a spreadsheet. A simplified example of the revenue side (ignoring interconnect) for a single type of call (say calls made during the peak period to the national PSTN) is shown in Table 8.7. However, the difficulty is in obtaining the estimates of the total subscriber numbers, minutes of usage, and call destinations. For a network already in existence, these can be extrapolated from existing records. For a new network, knowledge of other networks in the same or similar countries can provide valuable input. Nevertheless, this is essentially a forecasting role, with the skill set that such a role entails. Typically, the complete wireless professional would rely on the marketing department to derive such information based on:

Table 8.7
Example of a Spreadsheet for Predicted Revenue

Annual subscription/customer ($)	100
Fall in subscription/year	10%
Call minutes/subscriber/year	1,600
Cost/minute ($)	0.4
Rise in usage/year	10%
Fall in cost/minute each year	15%
Bad debts	5%

	Year				
	1998	1999	2000	2001	2002
Ttl subscribers ('000)	50	200	500	1,000	1,500
Ttl subscriptions ($m)	5	18	40	70	90
Call minutes (millions)	80	352	960	2,080	3,360
Cost/minute	0.40	0.34	0.29	0.25	0.21
Less bad debts	5%	5%	5%	5%	5%
Ttl call revenue ($m)	30.4	113.7	263.6	485.4	666.5
Ttl revenue ($m)	35.4	131.7	303.6	555.4	756.5

▶ Data from existing networks;

▶ Demographic information;

▶ Information about the wealth of the populace;

▶ Information about existing use of telephony;

▶ Predictions for GDP trends and population trends;

▶ Awareness of relevant changes such as an increase in home working;

▶ Guess work!

Predicting interconnect fees can be relatively simple since these are often published by the PTO, but the prediction of tariffs can be difficult. Few would doubt that tariffs would fall in the future, but estimating by how much is nearly impossible. An interesting exercise is to understand the minimum tariffs at which it is still profitable to operate the network and to assume that, at the worst case and in a price war, tariffs might at some point fall to this level.

In reality, considerably more complex spreadsheets than that shown in Table 8.7 would be required.

It is worth considering the interconnect arrangements in more detail since this is a subject of some complexity. Interconnect is payable whenever the call originates on a different network than that on which it terminates. The user in the originating network is charged for the call and then the originating network pays a fee to the terminating network for carrying the call. Different interconnect fees will be payable depending on the destination of the call. Interconnect revenue results from calls that are made to mobiles. Table 8.8 shows a simplified interconnect calculation for a single subscriber over the period of a year where the tariff is the amount charged to the mobile user for making the call, while the cost is the interconnect fee paid to, or received from, other operators.

In this example, the difference between peak and off-peak rates has not been shown nor has the interconnect arrangements relating to roaming mobiles (i.e., mobiles registered on a different network). The PSTN is the fixed network, while the *public land mobile networks* (PLMNs) are the mobile networks. The figure shows that considerably more fees are received for terminating incoming calls than are paid to other operators for terminating outgoing calls, and as a result the net interconnect revenue is significant. Note also the assumption that the mobile originates slightly more calls than it receives, which is typical for mobile subscribers.

Recently, interconnect fees for calls made from the PSTN to the mobile network have been a subject of much scrutiny in the EC. These fees are typically very high, leading to the cost of calling to a mobile being high. This results in a higher revenue for the mobile operator. Many suspect that the mobile operator will attempt to reduce the tariffs they charge their users since the users are well aware of these but increase the interconnect costs since few callers from the fixed network are aware of the cost of calling a mobile. In this manner, they can appear to be relatively inexpensive to their users. Mobile operators typically deny that this is the case and claim that the cost of terminating these calls is high; but since they charge more for terminating a call arriving from the PSTN than they do for a mobile-to-mobile call on the same network, which uses twice the resources, there is clearly some need for further investigation in this area. It seems likely that in the future, balanced and fair interconnect rates

Table 8.8
A Simplified Interconnect Calculation

| Call minutes per subscriber/year | 1,600 |
| Increase in call minutes | 10% |

	Percent	Tariff $/min	Cost $/min
Incoming calls			
From PSTN	24	—	0.42
From other PLMN	9	—	0.19
From own PLMN	7	—	—
International	7	—	0.08
Outgoing calls			
To PSTN	25	0.47	0.06
To other PLMN	11	0.28	0.22
To own PLMN	8	0.22	—
International	9	0.75	0.06

Interconnect revenue in $/subscriber/year	Year				
	1998	1999	2000	2001	2002
Incoming calls					
From PSTN	160	176	194	213	234
From other PLMN	28	31	34	37	41
From own PLMN	0	0	0	0	0
International	9	10	11	12	14
TOTAL	197	217	239	263	289
Outgoing calls					
To PSTN	22	24	27	30	33
To other PLMN	39	43	47	52	57
To own PLMN	0	0	0	0	0
International	8	9	10	11	12
TOTAL	69	76	84	92	102
Outgoing less incoming	128	141	155	170	187

Recently, interconnect fees for calls made from the PSTN to the mobile network have been a subject of much scrutiny in the EC. These fees are typically very high, leading to the cost of calling to a mobile being high. This results in a higher revenue for the mobile operator. Many suspect that the mobile operator will attempt to reduce the tariffs they charge their users since the users are well aware of these but increase the interconnect costs since few callers from the fixed network are aware of the cost of calling a mobile. In this manner, they can appear to be relatively inexpensive to their users. Mobile operators typically deny that this is the case and claim that the cost of terminating these calls is high; but since they charge more for terminating a call arriving from the PSTN than they do for a mobile-to-mobile call on the same network, which uses twice the resources, there is clearly some need for further investigation in this area. It seems likely that in the future, balanced and fair interconnect rates will become increasingly prevalent as regulatory pressure forces mobile operators to move in this direction.

8.2.5 Financing

Financing is the process of deferring payment in the early years of the network. It is effectively the borrowing of money. This is often required by mobile radio operators since they must make an investment in the network before they recoup that investment through call charges. Although at the highest level this would seem to be simply a case of borrowing money, there can be considerable complexity in the manner in which financing is achieved.

The mobile radio operator is looking for a source of borrowing at the lowest possible rates of interest. The lenders are looking for a way of making a return on their money at the lowest possible risk. The complexity arises in meeting the objectives of both parties. Often, investment in a mobile radio operator will be looked upon as relatively high risk since the timescales before repayment can be long and, particularly for the third and fourth mobile radio operators in a country, there is little evidence that they will be successful. Banks may not be willing to lend money against such a high perceived risk.

In outline, the options available for borrowing are:

‣ Self-funded from reserves within the group;

- ▶ Direct funding from banks or other financial institutions;

- ▶ Vendor financing from the equipment manufacturer;

- ▶ Shareholder funding.

The first option is only open to large companies who have substantial cash reserves and is becoming increasingly rare. It has the advantage of the operator not having to take any action to secure the funding.

The second option of borrowing from financial institutions is also relatively straightforward. There is typically no loss of control of the company and repayment schedules can be set to meet the requirements of both parties. Most banks will see a mobile telecommunications project as a high-risk investment and will only want to take a limited stake in order to reduce their exposure; as a result, it is typically not possible to generate all the funding required from bank lending.

There have been many vendor-financed sales in recent years where the manufacturer essentially provides their equipment for free and then is paid back as the network becomes profitable. In principle, there should be little need for vendor financing since there exist many other sources of financing. Vendor financing is normally used where the risk is considered so high that no other party is prepared to make the investment at a reasonable rate of interest. It is worth asking why, if the loan is so risky, the vendor might be prepared to take it on. There are a number of reasons for this.

- ▶ The vendor sees this as the only way to sell equipment on which they will make a profit—however, if the company collapses they will not realize their profit.

- ▶ The vendor is in a better position to judge the risk than the bank because they typically have a better understanding of the industry.

- ▶ To some extent, if the operator fails the vendor can remove their equipment and sell it to a different operator, effectively reclaiming part of their loan; whereas banks will typically not be able to sell any equipment that they reclaim.

Basically, the manufacturer is expected to provide funding by deferring the payments on the equipment or alternatively the equipment can be used as a means to purchase equity (shares) in the company and the manufacturer then becomes a shareholder in the operator. Manufacturers reluctantly agree to such projects if it is the only way in which they will sell equipment. However, vendors themselves can only borrow certain amounts of capital and cannot finance all the projects for which they are asked. Many vendors will have problems with high levels of financing because this will result in poor sales figures for the year because the financed sales cannot be realized on the books until a cash payment is received. Hence, an operator may have to select a vendor other than that selling the most appropriate technology if vendor financing is required. For this reason, a number of institutions such as the *European Bank for Reconstruction and Development* (EBRD) are rather wary of vendor-financed projects, preferring to select the technically most appropriate equipment rather than the operator who can provide the most funding.

Shareholder funding generates revenue through the use of an *initial public offering* (IPO) that places the company on the stock market. Typically, it is not possible to perform an IPO until the network has already started to be rolled out and there is some revenue from subscribers to allow investors to better understand the viability of the company. As a result, shareholder finance can only be used during the later stages of a network (typically not until at least one year from license award and often two). Shareholder funding has the advantage that there is little need for repayment schedules to be met but the disadvantage that some of the control of the company passes to the shareholders.

For a more detailed introduction to financing methods, readers are referred to [4].

8.2.6 Summary

Having calculated the total expected capital, ongoing costs, and the incoming revenue, the next step is to compare the two in order to determine whether the selected balance of coverage, service, and tariffs meets the business requirements. There are many ways of examining the information (again, readers who desire a full introduction to project financial analysis should consult [4]). The first step is to compare the total expenditure with the total income. This is shown for the simple example in Table 8.9.

The spreadsheet shows an unusually profitable network (deliberately made so to keep the example simple); in practice, much longer break-

Table 8.9
Example of a Spreadsheet Showing Summary Financial Statistics

Basic assumptions	
Interest payable on loans	10%

	Year				
	1998	1999	2000	2001	2002
Handset subsidy ($m)	29	82	155	246	233
Capital spend ($m)	11	29	58	97	97
Ongoing spend ($m)	4	14	35	68	101
Revenue ($m)	29	107	248	453	616
Profit ($m)	−15	−18	−1	41	183
Borrowing ($m)	−15	−35	−41	−3	179
Interest payable ($m)	−1	−3	−4	−0.3	0
Bank balance ($m)	−17	−39	−45	−4	179

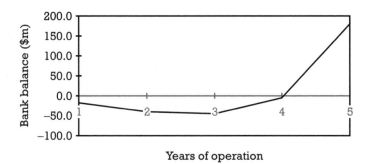

Figure 8.2 Cumulative profit by year.

even times are likely. Nevertheless, it adequately and simply demonstrates some of the key financial analysis that is required.

One key statistic immediately visible is the total funding requirement. The network will need to borrow over $45M during 2000 before starting to repay the loans in following years. This is a key input into the funding analysis.

Another key figure is the time to break even. This is shown in Figure 8.2. This shows that the break-even point was achieved approximately four years after the network was constructed.

The financial measures quoted most often are the *net present value* (NPV) and the *internal rate of return* (IRR). These are derived from the cash flow for each year, which is in turn derived as the revenue less the expenditure for that year.

One of the most appropriate ways to evaluate a project is to consider the NPV, which looks at the total outgoings and incomings over the length of the project, weighting money spent or received during future years by a factor accounting for the cost of capital over this period. Management texts agree that this is the most appropriate way to evaluate a project. In principle, any project with a positive NPV should be undertaken.

Another investment measure that is widely used is the IRR, which effectively provides a measure of the interest that is gained on the money invested in the project. If the IRR is higher than the cost of borrowing the money, or the return that can be achieved from investing it elsewhere, then the project should go ahead. IRR is widely disliked in management texts because, were there two mutually exclusive projects, one with an IRR of 20% and an investment of $100M and the second with an IRR of 30% and an investment of $20M, there is clearly a much larger absolute amount of profit to be made on the first project simply because of the larger sums involved. Because IRR does not consider the total sum invested, only the equivalent interest gained, a company would,

Table 8.10
Example of a Spreadsheet Showing NPV Calculations

Assumption	Discount factor	10%

	Year				
	1998	1999	2000	2001	2002
Discount	1	0.9	0.8	0.7	0.6
Total outgoings ($m)	44	126	249	411	432
Discounted ($m)	44	113	202	300	283
Total revenue ($m)	29	107	248	453	616
Discounted ($m)	29	96	200	330	404
Total ($m)	−15	−16	−1	30	120
Cumulative ($m)	−15	−32	−33	−3	117

incorrectly, select the second investment. Hence, NPV is considered a more appropriate investment mechanism. IRR is, in fact, closely related to NPV analysis—the IRR is nothing other than the discount rate, which, if applied to an NPV analysis, would yield an NPV of exactly zero.

In order to calculate the NPV, the preceding table needs to be rearranged slightly to look at inflows and outflows and cost of capital.

Table 8.10 shows that when considered as a 5-year project (in practice, most mobile networks will be considered over 7 to 10 years but 5 was selected here for simplicity) the project has a NPV of $117M. In principle, any project with a positive NPV should be undertaken.

In practice, NPV can be calculated most simply using the "NPV" function in most spreadsheets.

References

[1]   Allen, M., and D. Myddelton, *Essential Management Accounting*, New York: Prentice-Hall, 1992.

[2]   Dyson, J., *Accounting for Non-Accounting Students*, London: Pitman, 1991.

[3]   Emmanuel, C., D. Otley, and K. Merchant, *Accounting for Management Control*, London: Chapman and Hall, 1990.

[4]   McLaney, E., *Business Finance for Decision Makers*, London: Pitman, 1991.

CHAPTER

9

Contents

Operating a mobile radio network

Business has only two basic functions—marketing and innovation.

Peter Drucker

9.1 Introduction

Operating a cellular network consists of both keeping the network working and monitoring usage so that areas where congestion might occur can be predicted. Where congestion is expected, network capacity enhancement will be required. It is also important to understand marketing policies since changes in areas such as tariffs can lead to a sudden dramatic increase in usage. The complete wireless professional should understand:

▶ How statistics are generated and processed regarding the daily operation of the network;

▶ The implication of changes in marketing on the network design;

▶ Ways to enhance the capacity of a radio-based network once congestion is predicted.

9.2 Monitoring the network

It is important to continually monitor the operation of the network in order to enable:

▶ Delivering statistics about the network so that the need for additional resources can be monitored;

▶ Changing network configuration;

▶ Monitoring the correct working of the network and raising alarms when failures occur;

▶ Allowing new subscribers to be entered onto the network database and old ones deleted;

▶ Monitoring the network for fraudulent use;

▶ Generating bills for users;

▶ Providing customer care by monitoring usage, providing new features, and managing supplementary services.

Delivering network statistics   Network statistics are vital to ensure that the quality of service parameters for the network are being met. The key statistics include:

▶ The percentage of time for which all the channels in a cell are in use (note that the network cannot directly return a blocking statistic because, by definition, it does not know about blocked calls; however, if all the channels in a cell are in use, it is a good indication that blocking might exist);

▶ The percentage of calls that have been dropped, which provides information on quality of signal strength, quality of handover, and possible interference problems;

▶ The voice quality as measured in GSM by a function termed RX-QUAL (received quality), which estimates the voice quality based on the number of received blocks in error;

▶ Network downtime resulting from a failure of a part of the network.

With these statistics, the operator is able to determine the changes that are required to the network in order to reach targets. For example, where cell sites are congested, more channels can be added or microcells installed in order to increase the capacity. Where dropping is problematic, a team can be sent to investigate whether the problem is related to interference or coverage.

The biggest problem with network statistics is making sense of the vast amount of data that the network can provide, running to gigabytes of information per day. Sophisticated software packages are now available running on a central data collection unit that sifts through the data looking for patterns and evidence of specific problems.

In the future, manufacturers are predicting that, using statistical information, networks will be able to self-optimize, deciding on the frequency assignment and the cell neighbors for the purpose of handover. To do this, the measurements made by the mobiles on the signal strength in the surrounding cells will be passed back to a central database, as well as used to make a handover decision. By analyzing the signal strength from surrounding cells as perceived by hundreds of mobiles in a given cell, an accurate picture of interference and neighbor cells can be built up. It is thought that this sort of application will be deployed from 1999 onward.

Changing network configuration   When it is determined that a network needs to be modified in order to improve the quality of service for the subscribers, it is typically necessary to change the network configuration. For example, if microcells are to be added in an area, it may be necessary to remove some frequencies from the macrocells and to replan the macrocell frequency assignment. In a typical network, large-scale configuration changes are made three to four times a year. All the base stations must be changed within a few minutes of each other since during the change process there is likely to be poor network quality because some base stations are using the old parameters and some the new. The only way to achieve this coordinated change is via the use of a central device able to run a batch file with all the change instructions in place and send all the appropriate commands to the different base stations. This function is performed by the *operations and maintenance center* (OMC).

Monitoring the correct working of the network Subscribers will judge a mobile radio network based on the quality of service that they receive. This is composed of a combination of blocking, lack of coverage, and network failure time. Any network that has long periods of down-time will lose revenue both directly through the calls that cannot be carried while the network is down and indirectly through the loss of subscribers who find the quality of service too low. To help overcome long periods of down time it is essential that the functions of the network are continuously monitored. This is achieved through the function of the OMC, which provides a centralized terminal on which all the functions of the network can be monitored. In a system such as GSM, the interface to the OMC has been standardized so that components from different manufacturers, such as BTSs, can all interwork with the same OMC. In GSM, each of the base stations reports back on its correct operation to the BSC that controls it. The BSC is then responsible for reporting back to the network management system both about its own functionality and the functionality of all the BTSs under its control. This reporting takes place along the A-interface link from BSC to MSC and then along the E-interface link from MSC to MSC before finally arriving at the OMC.

Adding and deleting subscribers When a potential new subscriber enters a shop and buys a phone, he or she now expects that phone to be working by the time he or she returns home. In order to activate the phone it is necessary to register it onto the HLR and provide it with an appropriate service record. Billing records also need to be generated, although they do not need to be in place quite as quickly as the HLR registration. In order to achieve this, shops will often be connected to a distributed computing system (e.g., an Intranet) that allows them to send electronic registration messages into the network that can be forwarded to the HLR. Obviously, a similar process is required to delete subscribers.

Monitoring fraud Fraud has been an increasing problem for network operators for some time. Although the process of cloning analog phones is reducing as more subscribers use digital phones, there are still problems with theft of phones and with complex call-forwarding call schemes. Generally, once a phone is stolen or forwarded the fraudsters know that they have very little time to make the most of the phone before the network operator catches up with them. As a result, they tend to use the phone almost continuously and for international calls. Network operators try to catch such behavior by monitoring the usage of each phone. If it suddenly increases substantially, particularly with international calls, the fraud-monitoring system running on the billing platform may shut the phone down. Clearly, there will be difficulties in finding a balance between rapidly stopping fraudulent use and allowing users to change their calling habits, for example, as the result of a new job.

Generating bills   Within GSM, each call generates an electronic billing record. This is forwarded to the billing system in the home network where it is held against the subscriber generating the call. At the end of the accounting period (typically a month), the billing system automatically produces a printed bill detailing all the calls made and the charges payable. Advanced billing systems even automatically insert the bill into an envelope.

Operators are demanding increasing flexibility from the billing system to deal with a wide range of tariffs and special offers, for example. Consider that one operator has a range of different tariffs and promises that each month the billing system will select the tariff most suited to that user and bill them accordingly. As packet data systems arrive, billing will become even more complex. As a result, it will not be unusual for expenditures on billing systems to reach many tens of millions of dollars and for lead times on developments of the platform to be the key constraint in launching a new service.

Providing customer care   Customer care systems might provide the customer with the help and information that they require to make better use of the mobile phone. Examples include the provision of football results, share prices, and similar data. To provide this service, a bridge between the information providers and the mobile network is required.

When trying to add additional customer care features, the value of having a system conforming to an open standard rather than being proprietary can be very high. For proprietary systems, the operator is reliant on the supplier to add any required features to the equipment and the supplier may not be prepared to place the development of this feature high on their priority list since they are now in a monopoly supply position to this user. When an open standard has been procured, it is often possible for third parties to provide software or hardware providing the required functionality. A good example of this is within GSM where an IN peripheral to the MSC has been defined, known as CAMEL (see

Section 11.3). This allows third parties to provide software that can alter the manner in which the switch from a different vendor operates. Because the third party is under competitive pressure from other companies, the new features are typically provided more quickly, at less cost, and often requiring less effort from the operator.

9.3 Tariff policies and their implications

Tariff policies and network capacity are linked in a complex manner. It might be imagined that if the tariff is reduced, then users will make more calls with their mobiles. As a result, the capacity required from the network will be increased. However, predicting the manner in which usage will vary with traffic does not appear to be possible in any accurate fashion at the moment. With this relationship in mind, a mobile radio operator might consider the required capacity for their network in a number of different ways.

1. Capacity could only be provided as required by actual traffic growth.

2. Capacity could be provided in advance of predicted growth to provide a constant buffer against unexpected growth surges.

3. Capacity could be dramatically increased in order to be able to reduce tariffs and attempt to gain significant fixed/mobile substitution traffic and market share.

Fixed/mobile substitution is a key issue in the future progress for cellular. In developed countries, most of the telephony traffic (around 98% by volume) is still carried over fixed networks. However, some cellular subscribers are now starting to see their mobile phone as their primary means of communications and are making calls on the mobile phone that would previously have been made on the fixed line. At present, the key barriers to this continuing are the higher cost of mobile calls, the lower voice quality, and the lack of reliability in terms of high blocking and dropping rates. However, there have been many attempts to overcome these problems using reduced tariffs for calls made in home cells and similar. A network strategy that dramatically increases capacity and reduces prices could result in a massive increase in cellular traffic.[1]

The strategy selected has a number of key impacts on the capacity enhancement technique that should be adopted (and hence on the network plan developed by the engineer). For example, in case (1) it is only necessary to determine the least expensive means to provide the additional capacity. With cases (2) and (3) it is necessary to consider a number of interrelated factors such as:

▶ How the number of subscribers will grow in the coming years given the tariffing policy;

▶ How the traffic per subscriber will change given the tariffing policy, subscriber mix, and extent of fixed/mobile substitution, which itself will be influenced by coverage and blocking;

▶ How the action of competitors will impact on the tariffing policy and the total subscriber numbers, including areas such as fixed/mobile integration and mobile computing;

▶ Leading from these, how the revenue per subscriber will change as a result of all these interrelated policies.

An intelligent operator might attempt to determine the tariff level at which their revenue would be maximized, taking into account the total revenue at any point and the cost of providing the necessary network capacity. This calculation can be performed using a spreadsheet model if some assumptions about the manner in which the usage would vary with the traffic could be made. The complete wireless professional should be able to work with the marketing and financial staff to determine where the optimum point of network operation might lie.

---

1. A strategy along these lines was adopted by One-2-One in the United Kingdom, who offered free calls after 7 P.M. The result was a massive increase in traffic, with One-2-One users generating over three times the average traffic on the other cellular networks. Unfortunately, One-2-One was unable to economically expand their network to handle this traffic and decided to discontinue the tariff. However, this clearly shows that with a reduced tariff, dramatic increases in traffic volumes are possible.

9.4   Capacity enhancement

9.4.1   Introduction

In the case that demand for service exceeds supply and the blocking increases it is necessary to enhance the capacity. This section looks at means whereby the capacity can be enhanced. It considers only GSM—typically other cellular networks are not yet in a position where they need capacity enhancement, and hence the options for capacity enhancement are much more limited. The logical approach to understanding capacity enhancement described here, however, should be equally useful when considering other technologies.

9.4.2   The available capacity enhancement techniques

We first seek to understand the underlying theory. The capacity of a cell in terms of number of radio channels is given by

$$m = \frac{B_t}{B_c K} \tag{9.1}$$

where B_t is the total bandwidth (spectrum) assigned to the operator, B_c is the bandwidth required per call, and K is the cluster size or reuse factor. To enhance the capacity it is necessary to change one of the terms in this equation. Each of the possible means is now considered.

Increase the total bandwidth   Many operators are not able to gain additional spectrum. However, recently there has been a trend toward awarding dual-band spectrum. *Dual-band operation* is seen by many as a means to increase their network capacity.

Reduce the bandwidth per call   There is only one way to achieve this, which is through the use of the *half-rate coder*. However, operators have generally dismissed this coder as having inadequate voice quality and it will not be considered here.[2]

---

2. Actually, the picture here is still quite unclear. In 1997, all operators were saying that half-rate would not be used because of poor quality. By 1998, some of these operators were looking at half-rate deployments. Many operators are not publicly announcing their intentions because of the politics of requesting further dual-band or UMTS spectrum. However, as mentioned in Section 2.4.2, the most likely scenario appears to be operators waiting for the advanced multirate coder before moving to lower speech coding rates.

Decrease the cluster size   To decrease the cluster size it is necessary either to decrease the interference transmitted into cochannel cells using *transmit power control, discontinuous transmission, multiple reuse patterns,* or *concentric cells* or to increase the tolerance to interference using techniques such as *frequency hopping* and the *enhanced full-rate coder.*

Use more cells   Since the equation describes the number of channels per cell, simply increasing the number of cells will increase the capacity. This could be through *cell splitting, sectorization, microcells* or *picocells,* and *underlay/overlay cells.*

Changing one of the terms in this equation is the only means whereby it is possible to increase the capacity of a cellular system. Measures from all families could be used at the same time if required. The capacity gain from each technique is now considered in a little more detail.

9.4.3   Dual-band operation

The maximum gain in the number of radio channels is simply given by the percentage increase in spectrum (e.g., if the 1,800-MHz assignment is the same size as the 900-MHz assignment, then the number of radio channels is increased by 100%). This leads to even greater gains in traffic capacity due to the trunking efficiency achieved, but here, for simplicity, gains will only be discussed in terms of the increase in the number of radio channels.[3] The maximum capacity gains can only be achieved if there is at least as high a percentage of dual-band mobiles in use as the percentage of the total assignment that the 1,800-MHz spectrum forms (50% in this example). These gains also require algorithms to ensure that mobiles are reasonably evenly distributed across the two bands, but in principle this is not too difficult.

9.4.4   Techniques affecting the cluster size

In a cellular system, the maximum capacity is achieved when the interference is distributed evenly across all mobiles. This is a fundamental design

---

3.   To illustrate the trunking gains, assume that the operator had 2 × 12MHz at both 900 and 1,800-MHz frequency bands. Assuming an average cluster size of 15, this results in 4 carriers, or around 30 voice channels in each band. At a 2% probability of blocking, 30 channels provides 21.5E whereas 60 voice channels provides 49E; hence, a doubling in spectrum provides a capacity gain of 2.27 as a result of the trunking gain.

concept in mobile radio systems. Imagine cell A using 10 frequencies and cell B, some distance away, also using the same 10 frequencies. A system designer wants to be sure that the interference on each of the frequencies is sufficiently low that reliable communications can be made. Put together, the mobiles in cell A are going to generate a certain amount of radio signal. If this is not evenly divided across all frequencies, then one mobile in cell B will receive more interference than the others. To make sure that this mobile can communicate correctly, the cell needs to be far enough away so that this interference level is acceptable. However, this implies that on the other nine frequencies, the interference level is even lower. This is great for those mobiles who get a better quality, but it means that on those nine frequencies, the frequencies could have been reused closer to cell A, resulting in a lower cluster size and greater spectrum efficiency. Only if the interference is evenly distributed across all the mobiles can the minimum cluster size be adopted and the maximum capacity achieved. Choices made by operators about cell size and handover strategy can also have a significant effect on interference.

An operator can do two things. One is to adopt one of two ways to distribute interference more evenly—either frequency hopping or a form of dynamic channel allocation. The second is to reduce the interference generated using discontinuous transmission and power control. These are now considered in more detail.

9.4.4.1   Distributing interference evenly

Frequency hopping   To frequency hop, a mobile moves from frequency to frequency while it is transmitting. Other mobiles in the cell are also hopping from frequency to frequency and the sequence that each uses is designed so that no two mobiles in the same cell ever use the same frequency at the same time.

For any two mobiles using the same frequencies in cells that are some distance apart, in some cases the mobiles will be as close to each other as possible on the near edge of their cells, whereas in other cases the mobiles are much further apart than in the worst case and hence will suffer less interference. When designing a system without frequency hopping, the designer must assume that the worst case will occur. Hence, the designer must space cells using the same frequencies further apart than if the average case was assumed. This results in an increased cluster size and hence a reduction in the overall system capacity. This is inefficient because in most cases, the worst case will not happen. When using frequency hopping, the designer can assume that the average interference will be experienced by all mobiles and hence bring cochannel cells closer together.

There are two types of frequency hopping: synthesizer and baseband frequency hopping. Synthesizer hopping is a technique where mobiles hop over many frequencies (up to 64). In baseband hopping, the carrier frequencies transmitted by the base station do not change and the mobile just moves from carrier to carrier. Because there may only be two or three carriers per base station, baseband hopping results in a much lower number of frequencies to hop over and hence less randomization of interference. Synthesizer hopping should be used wherever possible.

Theory suggests that, depending on the design of the underlying network and the existing cluster size, gains of around 50 to 60% in the number of radio channels (and hence greater gains in the traffic carried due to trunking efficiencies) can be achieved using frequency hopping. Some operators have experienced larger gains than this, but these are probably due to a network that was somewhat inefficient before the capacity enhancement work started.

Dynamic channel allocation   The other way to evenly distribute the interference is to use DCA, where a cell does not have any frequencies permanently assigned to it, but when a user requires a channel, the base station selects that with the lowest interference. This removes the need for frequency hopping since all channels can now be selected with controlled levels of interference. DCA is not available within GSM, but a simpler approach, known as concentric cells (also known as *intelligent underlay/overlay* (IUO)) can be used that provides some of the gains of DCA. Here the mobiles near the center of the cell use one transceiver (TRX) while those on the edge use another. Those near the center are transmitting with lower power and the "edge of the cell" generated by this TRX is well within the edge of the cell generated by the other TRX. Hence, it will be possible to reuse the frequencies in the inner cell much closer to the first base station than those used in the outer cell. The inner cell will have a smaller cluster size and hence more capacity is gained. This technique reduces the need to randomize interference because the interferers in any particular cell are now more alike in their position in the cell and transmitter power. Its key advantage is in reducing the cases where spectrum is wasted because two cochannel mobiles are close to their respective base stations and hence generating little interference.

The maximum theoretical gain from DCA also depends on the cluster size in the network before enhancement, but is typically around 50%. For the concentric cell approach to achieve this, an infinite number of layers of concentric cells would be required, so clearly the gains from concentric cells must be below 50%.

Both frequency hopping and DCA achieve an even distribution of interference, and both can provide similar gains of the order of 50%. Since both achieve the same thing, there is little merit in using both together. The only advantage would be to allow some frequencies to be used that had too high an interference to offer acceptable voice quality. As long as these are only hopped onto irregularly the reduction in quality may be acceptable. However, this is a dangerous arrangement since a reduction in signal strength could drop the call and this tactic should probably be avoided. Hence, the maximum gains that could be achieved with this set of techniques are probably around a 50% increase in traffic channels. Because concentric cells are not a perfect form of DCA, using frequency hopping as well does bring some gains over concentric cells alone, but these gains are typically not significantly greater than what could have been achieved with frequency hopping alone.

Multiple reuse patterns are very similar to concentric cells. The concept is that the first TRX in the cell operates on a cluster size of, say 15; the second on a 12 cell cluster; and the third on a 9 cell cluster. Quality is then higher on the first and second TRX than the third, but because of the reduced cluster size, the frequencies are reused more in the network. Users can hop across all the carriers, resulting in an average quality level, or can be directed away from the low-quality carriers by intracell handover in the case that their call quality is suffering unduly. The result may be something like a concentric cell—only those mobiles near the cell center will be left on the low-quality carriers because they will have a better signal strength being closer to the transmitter.

In principle, multiple reuse patterns are not necessary—a reduction in the cluster size on all TRXs coupled with frequency hopping will provide the best results. However, the TRX on which the system information is provided—the so-called BCCH carrier—cannot hop for design reasons, so it is necessary to use a larger cluster size for this carrier. Hence, the optimal design is typically a two-tier reuse pattern with a reuse of say 15 for the BCCH carrier and perhaps 9 for all the other carriers, all of which are using frequency hopping.

9.4.4.2 Interference reduction

Discontinuous transmission and transmit power control reduce the interference levels since the mobile and base stations now transmit for less time and with less power. Typically, these techniques only bring benefits when coupled with frequency hopping that will spread the reduction in interference around all the mobiles, helping to reduce the cluster size. The gains here are difficult to predict and depend on the existing cell planning, handover strategy, and other capacity enhancement techniques in use.

In summary, the capacity enhancements that can be achieved using cluster size reduction techniques will depend on the underlying network, but up to 50% can be achieved by evenly distributing the interference, and perhaps a further 50% by reducing the interference.

9.4.5 Using more cells

Most operators have already embarked on the path of sectorization and cell splitting. Actually, sectorization does not provide the capacity gains that might be expected. Because the cell radius is unchanged, the effective transmitted power (the EIRP) remains unchanged and hence the reuse distance remains unchanged. However, each sector needs a different frequency, so the result is that the cluster size after sectorization is greater than the cluster size before. Worse, the lower number of channels in each sector compared with those in the cell reduces trunking efficiency. In practice, careful use of antenna directionality can provide some gains, perhaps up to a 100% increase, but sectorization does not multiply the capacity by the number of sectors introduced as might initially be thought. Its widespread use is due to its cost-effectiveness since additional cell sites are not required.

Microcells, as the name suggests, are cells that are significantly smaller than the current cells used within a network. For the network operator, the key issue is how many and where microcells are to be deployed. There are two main strategies:

1. Hot spot coverage, where isolated microcells are placed in areas where there is a lot of radio traffic (e.g., train stations);

2. Contiguous coverage of a region where microcells are used to cover at least all the main streets within a given region.

The problem with the former is that it is often difficult to locate traffic hot spots; they may only be temporary and may expand or move so that the original microcell placement becomes inappropriate. However, because only a small number of microcells are deployed, the costs are relatively low. Contiguous coverage overcomes the problems of locating hot spots by providing coverage throughout the area. However, some microcells may be minimally used, resulting in a more expensive deployment than need be the case.

In practice, an approach somewhere between the two is typically adopted, where high-density regions are identified (typically cells where high blocking takes place) and all the main streets within the coverage of that cell are covered by microcells. This removes the need to locate the hot spots to the accuracy of a microcell but ensures only those parts of the city with relatively high traffic levels are covered.

The capacity provided by microcells is simple to calculate—typically they have two carriers and, hence, provide around 15 voice channels. Frequencies in a microcell cluster can be reused with a cluster size as low as 4 due to the blocking effect of the buildings on the interfering signal.

Underlay/overlay systems are basically cellular networks where the macrocells have been left in place while the microcells were deployed. The aim of the operator is to carry most of the traffic on the microcells, but if there is insufficient coverage, or if the mobile is moving rapidly and would require numerous handovers in a microcellular network, then they are placed on the macrocellular network. They do not increase capacity significantly but make microcells more workable.

9.4.6 Which capacity enhancement techniques should be used when?

Of course, the most appropriate capacity techniques to use depends on the capacity enhancement required and on the state of the underlying network. In this analysis it has been assumed that urban cells are already sectored and are approaching the minimum radius (around 400m) practically possible for macrocells. Certainly, if there is dual-band spectrum available, then it should be used (operators will typically never refuse more spectrum). The costs of using this spectrum are mostly involved in getting dual-band mobiles into the subscriber base, but as these mobiles become more widespread, this will tend to happen in any case over time.

However, for most operators this will only be a 50 to 100% capacity enhancement—worth having but not sufficient to solve the capacity requirements for all time.

The next obvious step is frequency hopping. This is within the GSM standard and the mobiles are already capable of hopping. By hopping all but the BCCH frequencies and reducing the cluster size on all the hopped carriers, improvements in capacity of perhaps 50% can be achieved at relatively little cost. Multiple reuse patterns or concentric cells could be added in at this point, but after frequency hopping has been deployed the gains from these other techniques are probably minimal and the effort required to implement the techniques may be out of proportion to the gains achieved. Only in the case of a badly designed original network might they offer a way out of existing interference problems.

After this, the only option is the addition of more cells. Only cell splitting and microcells will provide significant capacity gains and in a well-developed network, cell splitting is typically no longer possible. Microcells can provide huge capacity gains—easily 500% or more in city centers, but compared to the other techniques discussed earlier are relatively expensive, requiring new cell sites and infrastructure—hence, the reason to use them as the last capacity enhancement technique.

So the approach to adopt for capacity enhancement is relatively straightforward:

▶ Deploy dual-band spectrum if available.

▶ Implement frequency hopping along with discontinuous transmission and power control.

▶ Deploy sufficient microcells to provide the remainder of the capacity required.

Contents

Large users of mobile radio networks

That which seems the height of absurdity in one generation often becomes the height of wisdom in another.

Adlai Stevenson

10.1 Introduction

Most of the users of mobile radios are the general public. These are individuals who have varying needs from their mobile phones, but for the majority the key needs are the ability to make a voice call wherever they are for a reasonable price and at a reasonable quality. Today's mobile phone systems coupled with a well-designed network are able to provide this requirement. It has often been predicted that in the future the public will require more data capabilities, perhaps the need for video-phones, and other advanced capabilities. The ability for future mobile phone systems to provide these facilities is discussed in Chapter 11.

At the moment, the most pressing additional requirements come from the existing large users of mobile telephony such as police and railways. These bodies use mobile communications as an integral part of conducting their safety-critical roles. As a result, they tend to put a much greater requirement on their mobile communications. Understanding the needs of these users is important for a number of reasons.

▶ The wireless professional may be working directly for these users and so will need to understand their requirements.

▶ Wireless professionals working for manufacturers will need to understand the requirements of all the potential user groups in order to correctly design their systems.

▶ Wireless professionals working for PAMR and cellular operators are likely to be involved in discussions with these bodies about the possibility of them moving away from self-provision and onto a shared network.

▶ The requirements of these large bodies may well become the requirements of small companies in the future.

In this section, two key sets of users are examined in detail, the police and the European railways. The police are representative of all the emergency services and perhaps have the most difficult requirements of all to meet. The European railways provide an interesting case because of their recent decision to move from PMR technology to cellular technology. An understanding of these two user groups will provide a good understanding of most large PMR users.

Some important points to remember when reading both of these descriptions.

▶ The perceived requirements are often heavily conditioned by the capabilities provided by the existing radio system; for example, most existing systems can only provide group calls (and not point-to-point calls), but this has become part of the way of working and now many users cannot perceive operation without these features. Some would argue that these features would not be required if the way of working was modified. However, although the engineer can question a user's requirements, if the user insists on these requirements, the engineer has to accept them even if they do not appear to be logical.

▶ Both these organizations have existing departments responsible for the provision of mobile communications. It is in the interest of the people in these departments to make the requirements as complex as possible to justify their existence. Outsourcing their communications requirements will seem to them a little like voting themselves out of a job because if communications are outsourced, their role will be much diminished.

▶ There are differences between different countries. In Europe, PMR users have the alternative of a robust and ubiquitous GSM system. In the United States there is limited national roaming and only analog AMPS networks that provide widespread coverage. Hence, in the United States there is a greater need to implement a self-owned system.

10.2 Railways

10.2.1 Introduction

Like almost all forms of transportation services, railways have been quick to embrace mobile communications as a means to provide greater efficiency and improved safety. It is difficult now to conceive of a complex railway system being able to operate satisfactorily without mobile radio communications. The current mobile systems, and the uses to which they are put, are discussed in more detail below. With the EC directive on the harmonization of the European High-Speed Network technology now in place, coupled with the rapid development and introduction of relatively advanced digital mobile radio systems, by the early 1990s the European railways were looking for a new digital radio standard.

The railways noted the benefits of digital communications over the current analog systems, including improved voice quality, greater security, more flexibility, and better spectrum efficiency, and their view was that railways could gain benefits through implementation of a digital radio system. There was also a growing realization that the use of proprietary systems was becoming increasingly expensive, both in the specification and the procurement phases, and that standardized solutions would be likely to be more cost effective.

In the early 1990s, the European railway body, the *Union Internationale des Chemins de Fer* (UIC) decided to specify a digital radio system to be implemented from 1998 onward to provide both pan-European operation and an advanced range of features currently not available in many national radio systems. It is instructive to examine this project, named EIRENE (European Integrated Radio Enhanced Network), and the manner in which the new system is to be implemented.

The overall scope of project EIRENE was to provide a specification for a radio system that could meet most requirements of the UIC European member organizations. The initial study involved determining what the needs were through examining the requirements documents compiled by the railways and by conducting a series of interviews with different railway authorities. These needs were then classified into sets of logical requirements and compared with the features offered by current and emerging communications systems, including those undergoing standardization. A detailed study was made of the most appropriate candidate systems leading to a recommendation to use GSM. Shortcomings were identified in the GSM capabilities, and modifications suggested to overcome these. The final part of the feasibility study involved producing a generic procurement specification for the proposed system.

10.2.2 Current railway communications within Europe

European railway authorities operate a variety of mobile radio systems. Indeed some countries have more than one system; British Rail, for example, introduced three different systems between 1984 and 1986, all of which were still in operation in 1998. Part of the reason for the number of systems is the wide range of applications. Communications to drivers in trains, between shunting teams or trackside workers, and among station staff are all necessary. Currently, different systems are being used to fulfill different roles.

Nevertheless, there is a fair degree of commonality among European track-to-train mobile radio communications systems, with many being based on a UIC standard known as 751-3, originally issued in the early

1970s. This document, or *fiche* as it is more commonly known, allows for simplex and duplex conversations to drivers, including the capability to make group calls and emergency calls. The system is based upon an analog FM standard with 25-kHz channel spacing operating around 460 MHz. Provision is made for tone signaling, including a number of single frequency tones to indicate channel-free, listening, warning, and pilot tones, for example. In addition to the ubiquitous voice communications, short data messages can be sent using two-tone FSK.

In the 751-3 standard a particular frequency is assigned to each geographical area, and when entering that area the mobile should automatically switch to the assigned frequency. The channel is essentially open in that anyone can transmit on the channel and anyone in the vicinity can listen, although in practice some form of privacy is made available through selective calling and lockout procedures.

In essence, the current railway system described in fiche 751-3 is a PMR-type system dedicated to a small range of applications. Unfortunately, owing to implementation differences, the 751-3 systems in different countries are often incompatible. Because of this, international trains, such as those destined to run between Britain, Belgium, and France via the Channel Tunnel, are being equipped with "chameleon" radios, capable of operating on each of the different radio systems. This chameleon radio required a common specification for all three countries that has proved to be a complex and time-consuming task. A new pan-European radio standard would clearly ease this problem in the future.

Because 751-3 has only been designed for a limited range of railway needs, it is difficult to provide the communications facilities for applications such as maintenance and on-train reservations without excessively wasting radio spectrum.

10.2.3 Railway requirements

Before specifying an appropriate pan-European mobile radio system, it was necessary for the railways to determine their requirements. Once these were established, the tradeoff between cost and utility for each item could be reviewed and the functionality needed from the radio system established. A well-designed system should be capable of meeting all current requirements and allowing some room for expansion, without being overspecified.

There is a surprisingly wide range of railway requirements, partly due to the fact that the railways both run trains and maintain infrastructure. The most obvious requirements are for controller-to-driver communications. Controllers are members of staff responsible for a particular section of track; they must perform signaling and communications functions to provide a smooth and safe passage for trains, and they must act as an emergency contact if problems develop. There may be a number of controllers for a particular stretch of track, each associated with various aspects of train control. Controllers need to be able to contact individual drivers and groups of drivers within their area. Messages from drivers also include emergency broadcasts that need to be rapidly relayed to all staff within the area.

In addition to controllers, there are other individuals who may wish to contact personnel on the train, the latter including drivers, guards, and catering staff. The call originators might include other drivers, staff at stations, and staff supervisors, for example, and message types may include voice, facsimile, data, text, and interactive computer messages. Such calls may in the future be international.

With a view to improving the network capacity and efficiency, remote control is envisaged by some railways. As well as the remote control of entire trains, this might include remote control by one driver of a multi-locomotive train or remote control of items such as points and level-crossing barriers. The radio system is a potential bearer for automatic train control communications, gradually removing the need for costly fixed signaling infrastructure.

A number of railway tasks are performed by teams that require local communications. Such tasks include shunting, trackside maintenance, and local communications at stations. Communications are typically over a short range and involve groups whose membership might change on a daily basis. In the case of trackside maintenance, the team may be working in a remote location where radio infrastructure has not been installed, and so direct mode (or set-to-set) communications, where one handheld radio communicates directly to another rather than via a base station, may be required.

In addition to these operational requirements, a number of passenger facilities can be envisaged, such as database access for timetable information and advance booking, passenger information concerning delays, and entertainment and news facilities. If trains are to compete successfully with aircraft on intra-European routes, such passenger facilities may be needed.

Finally, some other constraints were imposed by railway operations. These were:

▶ Communication to and from high-speed trains are required, potentially traveling at up to 500 km/h. Few radio systems are designed to cope with the rapidly changing radio channels caused by the Doppler shift associated with traveling at this speed.

▶ Station staff will normally want to call a train by its running number (a number corresponding to a particular timetable entry such as "the 15H55 from London to Paris") rather than by a telephone number. This is because the 15H55 from London may be a different physical locomotive each day and so possess a different telephone number, whereas the running number is relatively static. If the locomotive radio only had a fixed telephone number, then trying to remember or to determine the appropriate numbers as they change each day would be problematic. However, the maintenance depot will only know a locomotive by its engine number and will want to call it using this. Both means of addressing the locomotive must be possible. These issues are considered later in this section.

▶ Very rapid call setup is required, particularly in emergency situations.

10.2.4 PMR versus cellular

Given the wide range of railway requirements, the design of a mobile radio system that could meet them all was destined to be problematic. The decision was made to employ an ETSI-specified European standard mobile radio system in order to reduce the design cost and achieve economies of scale through markets outside the railway arena. Given the wish to employ a digital system, and the timescales for implementation of 1999, this essentially led to a choice between GSM, TETRA, or DECT, of which DECT was rapidly ruled out owing to its short transmission range and limited facilities. A comparison made in 1993 at the time when the railways were making their choice of GSM and TETRA with the current railway system is presented in Table 10.1.

Table 10.1
Comparison of a Range of Radio Systems Made During 1993

	751-3	GSM	TETRA
Technology	Analog FM	Digital TDMA	Digital TDMA
Frequency	460 MHz	900 MHz	400 MHz
Effective bandwidth per voice channel	25 kHz	25 kHz, 12.5 kHz with 1/2 rate codec	6.25 kHz
Call setup times	<1 sec	<10 sec	<1 sec
Data facilities	Basic	Yes	Yes
Group calls	Yes	No (except via conference call facility)	Yes
Direct mode	N/A	No	Under development
Wide area communications	No	Yes	Possible
Maximum speed	Not specified	250 km/h	200 km/h
Handover	Limited	Yes	Yes
Roaming	No	Yes	Yes
Authentication and encryption	No	Yes	Yes
Status	Available	Available	Standardization ongoing

The fundamental choice falls between a public subscriber cellular-based system, such as GSM, or a PMR-type system such as TETRA. The difference between the two system types is in the basic mode of communications offered. In cellular systems, one-to-one connectivity over a wide area is required, whereas in PMR, group and dispatcher calls (probably over a smaller area) are necessary. Historically, this has tended to mean that in a cellular system a pool of channels is available within each cell and when communications are required one channel is allocated to each user who wishes to pass a message to another user. In a PMR-type system, a channel is assigned to a group of users for the duration of their conversation upon which they can listen and transmit relatively freely. This fundamental difference concerning channel assignment makes cellular and PMR difficult to reconcile.

It became apparent that neither type of system was capable of meeting all the railway radio requirements. PMR-type systems have allowed driver-to-controller communications in the past, but the increasing volume of traffic; the requirements for one-to-one communications between a number of possible users on the train and the ground, even across international borders; and the desire for a fully integrated radio system for all aspects of railway work are making them less suitable. None of the systems available offered operation at train speeds of up to 500 km/h without modification, nor the indirect addressing required to facilitate railway communications.

A strong case could be made for either TETRA or GSM; what was required was a combination of both systems. TETRA offered a better fit to the railway requirements than GSM but presented a higher risk because it was not standardized (much less available) at the time. In contrast, GSM was rapidly becoming a global success. It was therefore decided by the railways that the risks involved in meeting the timescales were lower for a GSM system with modifications than for TETRA.

With the selection of GSM, a range of modifications to the GSM standard to fill the shortcomings identified in Table 10.1 were put in place. These form part of the GSM Phase 2+ features and are discussed in more detail in Chapter 13.

10.3 Police

10.3.1 Introduction

The police forces make widespread use of mobile communications. Most policemen on foot and police cars are equipped with mobile radio systems. Current systems, much like the railway systems, are typically open-channel PMR systems. These offer a number of advantages to the police in that with the open channel everyone in a particular area can hear what is going on and take appropriate action. A policeman can talk to the controller at the police station simply by pressing a button, and the simplicity of open channel working ensures that the call setup is less than a second, which may be important in some life-threatening situations.

However, these simple radios also have a number of problems. One of the key problems is an increasing lack of radio spectrum. If all communications need to be heard by all users in an area, then they need to be on the same frequency. One frequency can only accommodate a limited number of users, and with the demands of modern policing this limit has been met in many areas. Analog systems can be readily overheard by criminals, which is a significant problem. The police are also increasingly making use of mobile data for a wide range of tasks such as number plate checking, mobile fingerprint analysis, and the use of video information transmitted to officers. The police can envisage a dramatic increase in the future of mobile data applications. Designs of the police helmet of the future show a helmet with a built-in video camera and eyepiece, with video being transmitted back to the control center and images transmitted to the eyepiece of the helmet showing pictures of suspects and other relevant details. It is quite clear that modern police communications will need to be based around the most advanced digital communications systems possible.

10.3.2 Description of requirements

The requirements of the police force are wide-ranging and include:

▶ Point-to-point communications with encryption and a call setup time of less than 1 sec;

▶ Group and broadcast calls to a subset of users or to all users with a 1-sec setup time and the capability for acknowledgment of receipt;

▶ Data communications at as high a rate as possible, with both circuit-switched and packet data capabilities;

▶ Direct-mode communications so that policemen out of radio coverage can communicate with each other;

▶ Relay-mode communications where a mobile within the coverage area can relay communications to a mobile outside of the coverage area;

▶ Rugged handsets that will withstand everyday use;

▶ End-to-end encryption on all important calls;

▶ International interoperability so that police forces from more than one country can cooperate in cross-border operations;

▶ The ability to communicate with ambulance and firemen during emergencies;

▶ High capacity in city areas but good coverage in rural areas;

▶ A host of supplementary services, such as the capability for a controller to turn on the mobile of any of their members of staff without the mobile making any noise, allowing the controller to hear what is going on—this might be useful if a policeman is held hostage.

Meeting all these requirements is clearly going to be difficult. The difficulty arises in matching the simplicity and speed of the wide area emergency broadcast with the complexity of the point-to-point encrypted call.

10.3.3 Selection of radio system

The police forces throughout Europe, like the railways, performed a detailed assessment of all the available mobile radio standards against their exacting requirements. Unlike the railways, they concluded that GSM would not be able to meet their needs. It may appear strange that the police came to different conclusions from the railways because, in general, as can be seen from the preceding descriptions, the requirements of the police and the railways are actually quite similar. In the case of the police, the need for encryption is higher and there is a perceived need for a faster call setup (although many point out that since help will typically take many minutes to arrive, the difference between a 1- and a 2-sec call setup may not be one of life and death). Many are of the view that the police requirements could have broadly been met by a modified version of GSM in the same manner that the railway's requirements were being met. However, for a range of reasons, the police decided to adopt the emerging TETRA standard for their own communications needs. Because the standard was still in development at the time, the police were able to add the additional features to the standard that they required and ensure that the system was tailored to their needs.

As well as the selection of the radio standard, police forces in Europe are often facing other changes in the manner in which their mobile communications are provided. As can be envisaged from the list of requirements, the new radio system is likely to be substantially more expensive than the existing radio system. Most police forces are under pressure to reduce expenditure, and this pressure extends to the mobile radio system. The basic way to reduce the cost of a system of this sort is to share the system. Indeed, a moment's analysis shows that if the system is not shared among more users the cost will be very great.

In the United Kingdom there are around 95,000 policemen. A radio network to cover most of the country would require around 1,000 base stations. Hence, there will be on average 95 policemen per base station (many more in cities, many fewer in rural areas).

With a cellular network, there are around 3 million subscribers on the two main networks in the United Kingdom. The cellular operators have deployed around 2,000 base stations to provide coverage and capacity. Hence, there will be on average 1,500 subscribers per base station.

Clearly the cost of a shared public system will be dramatically lower than the cost of a dedicated police system. In order to reduce this cost burden, police forces have been looking at sharing their network with ambulances and fire crews. However, this will less than double the user base. In the United Kingdom, the attempt at sharing has been taken a step further with the instructions to the police that they must seek a private company to build their network, with the police paying an annual fee for the use of the network. The intention is that the winning company will be able to use shared resources such as the backbone network and many of the mast sites in order to reduce the cost of the system to the police. Further sharing could take place with TETRA users on PAMR networks, although the police are keen to retain their own radio spectrum so that they do not suffer from their network being blocked by other users during an emergency. Whether these approaches will be successful remains to be seen.[1]

10.4 Other emergency services

In this category fall the ambulance and fire services and some smaller organizations such as the coast guard and similar groups. Mostly the requirements of this group are a subset of those of the police. Because the groups are generally smaller, their requirements often get subsumed into police systems and they may end up sharing technology. For an ambulance service, the key requirements are:

‣ Dispatching (the sending of an ambulance to an incident);

‣ Status reporting (an ambulance telling the controller what it has found and what it is doing about it);

‣ Two-way text messaging (information and instructions);

---

1. At the time of writing only one consortia was interested in providing this service, with other competing consortia having withdrawn. This was raising concern that the remaining consortia was effectively in a monopoly supply position.

▶ *Automatic vehicle location* (AVL) in order to allow the controller to establish the location of all his or her resources.

Of these requirements, often the most difficult to meet on a standard radio system is AVL as this may require short data bursts every, say, 10 sec. Where the radio system only supports circuit-switched communications, then this can tend to be highly inefficient of resources. Supporting AVL efficiently requires a packet-based data communications system. A fire brigade has similar requirements but may also need:

▶ The ability to send building plans over the radio system, requiring high-speed data capabilities;

▶ The ability for the radio in a fire vehicle to act as a repeater to firefighters inside the building who may be out of coverage from the base station.

Most ambulance and fire services employ the same type of radio system as the police, and in Europe it is currently envisaged that all the emergency services will share a TETRA system.

10.5 Other users

There are many other users, as shown by Table 4.2, and covering them all would take some time. All have simpler needs than the railways and the emergency services discussed here.

Utilities such as gas, electricity, and water typically require good voice communications; the ability to send data; possibly the capability to remotely read meters, or to remotely interrogate customer databases; and good coverage along the pipelines and electricity pylons, for example. The latter is problematic as these often tend to be in rural areas where there is generally little coverage. Utilities require the capability to preempt calls in the case of an emergency, such as a gas explosion. It is typically the coverage and the preemption requirements that tend to result in the utilities deploying their own systems rather than sharing PAMR or cellular systems, but the increasing cost of self-deployment and the improvements in shared systems have meant that the utilities are slowly moving away from self-provision, though again they are reluctant to give up their own frequencies and systems.

The requirements for organizations such as delivery companies are relatively simple. They wish to send itineraries to vehicles at the start of the day and for the vehicles to be able to send proof of delivery notifications as they deliver their cargo. They would also like AVL. These requirements can all be met by a data-only system such as some of the public packet data networks in operation. However, the companies typically also like voice backup so that they can speak to the driver in the case that something unexpected happens. Such organizations now rarely have their own systems, typically using either PAMR or public data systems.

CHAPTER

11

Contents

Future mobile radio systems

Everything has been thought of before, but the problem is to think of it again.

J. W. von Goethe

11.1 Progress in radio systems

As discussed in Chapter 1, the first true cellular systems were the analog networks introduced in the mid-1980s. These are only now being phased out as the second generation of radio systems, the digital cellular technologies, are coming to the fore, some 12 to 15 years after the analog systems. The digital *(+)* systems are better, with better voice quality, better security, more services and data capabilities, and an increase in capacity. However, the increase is relatively small. Some operators have even claimed that they got more capacity from their analog systems than their *(−)* digital systems, although this generally seems to be a result of using the same base station

269

sites, which may not be ideally positioned for digital. A quick calculation shows that TACS-type systems have voice bandwidths of 25 kHz and cluster sizes of 21. GSM has the same voice bandwidth but an average cluster size of around 12 in most deployments. The capacity gain is therefore around 75% over 15 years. Given the rapid increase in the size of computer files that users might wish to send, it can be seen that mobile radio systems will have problems keeping up with the demands that users might place upon them.

As discussed in this chapter, the next generation of products will probably be well established to the extent that they rival second generation systems by around 2010, another 12 or so years away. It is too early to understand the capacity that will be offered, but an estimate of a further 75% increase is unlikely to be highly inaccurate. Services will be improved, but the average user may not notice a large difference. This chapter looks at the developments of third generation systems. Since the situation is currently changing rapidly, the interested reader will need to seek additional sources of information such as the ETSI web site [1].[1]

11.2   The third generation vision

Although, to some extent, the next generation of radio systems is simply an opportunity to update existing mobile radio systems as advances in technology and manufacturing have occurred, many of the standards bodies have attempted to set out their vision of what the standard should provide. There are four main parties involved: the ITU at an international level, the national regulators in Japan and Europe, and the CDMA development group, mostly based in the United States. Most parties agree that the goals of the next generation system should be "communication to everyone, everywhere." Communication in this instance might also include the provision of information. To achieve this goal it must provide all the facilities required for a wide range of disparate users including:

▶ Cellular users who require high voice quality and good coverage;

---

1. As an aside, learned journals cannot be recommended as an up-to-date source of information. It now takes longer to publish a paper in most learned journals than it does to publish a book!

▶ Users who currently deploy their own private systems (see Chapter 4) who require group and broadcast calls where all users can hear what one person is saying;

▶ Paging users who require a small terminal and good coverage;

▶ Cordless users who require excellent communications with high data rates when in the office or home;

▶ Satellite users who require truly worldwide coverage;

▶ Users in airplanes who currently have limited telephone availability (especially to receive calls);

▶ Data users ranging from telemetry to remote computer network access.

The system must also provide fixed mobile convergence such that users can get a similar service from a fixed or mobile connection, ideally using the same phone number. Current predictions as to the timescales of third generation systems vary slightly but typically the following is assumed:[2]

Standardization completed	1999
First product available	2002
Product widely available	2005

The current view is that GSM will evolve to provide third generation features, and hence it is necessary to obtain an understanding of GSM evolution in order to understand how the third generation will develop.

GSM evolution GSM is currently evolving to provide many but not all of the features of third generation systems [2]. Some of the key features being added to GSM as part of the Phase 2+ feature set include:

▶ *The advanced speech call items:* The *advanced speech call items* (ASCI) are defined as group and broadcast calls, and priority and preemption. Broadcast calls allow many users to listen to a single radio channel in a cell and for the call to be available in many cells. Group calls

---

2. These appear optimistic, and as discussed in Section 13.3, it would be no surprise to find the timescales slip by as much as 3 years.

allow users to respond after the call originator has finished speaking. Preemption allows a priority to be attached to the call and for other calls to be curtailed if there are no other resources available for a high-priority call.

▶ *A packet data service known as general packet radio service: General packet radio service* (GPRS) allows packet-switched communications. As can be seen from the discussion in previous chapters, this will provide increased efficiency in a range of applications, including Internet access. However, it requires a significant change to the GSM system. Alongside each MSC is a *GPRS support node* (GSN),which is basically a packet switch. There is then a separate packet network to support the GSNs and interconnect them. By sending packets on more than one time slot to the same user, data rates of up to 150 kbps can be achieved. There are many difficulties such as how a mobile listening on the GPRS channels for packets is able to receive paging messages—in this case, paging messages are also relayed on the GPRS channels, requiring careful integration between the GSN and MSC. It will be interesting to monitor the evolution of GPRS; as increasing volumes of data are transmitted over mobile phone networks it could be that GPRS becomes the dominant bearer in the GSM network. GPRS is seen by many operators as a stepping stone toward a UMTS system that will be predominantly packet based. Use of UMTS will help them understand the requirements, tariffs, and uses of packet data.

▶ *A higher data rate service known as high-speed circuit switched data: High-speed circuit-switched data* (HSCSD) provides a higher data rate than can be achieved at present by concatenating timeslots together. If 6 timeslots are used, then data rates of 64 kbps can be achieved. The only problems are the loss of capacity from giving one mobile a number of timeslots and the difficulty for the mobile in making measurements of surrounding cells when it has to listen to more downlink frames than normal.

▶ *Intelligent network capabilities, known under the misleading title of customized applications for mobile network enhanced logic: Customized applications for mobile network enhanced logic* (CAMEL) allows subscribers to make use of operator-specific services even when they have roamed to a different network. For example, an operator might offer its subscribers call forwarding depending on the time of day so that during the day calls are routed to the office and in the evening to the home. This is not part of the GSM standard but could be implemented with custom software in an MSC. However, when roaming, the subscriber would not be able to access such a service. CAMEL requires visited networks to send back event notes to the home network and then await further instructions. This leads to an architecture very similar to the IN in fixed networks, as described in Chapter 6.

▶ *The SIM toolkit:* The SIM toolkit is a package of measures that revolves around giving the SIM card some processing power and allowing it to interact with the mobile. By downloading applications onto the SIM card, new services can be provided to the user. For example, the operator may wish to offer a service whereby the user can book theater tickets using the mobile phone. The network would send a list of shows to the SIM that would then instruct the phone to display these to the user and return the answer to the SIM. The SIM would then instruct the phone to send a coded message back to the network so that the tickets can be ordered. The SIM toolkit coupled with CAMEL will go a long way toward allowing operators to provide personalized services to users wherever they are and whatever phone they are using.

▶ *The enhanced full-rate codec:* The *enhanced full-rate codec* (EFR) is a better voice coder, providing a higher speech quality by taking advantage of improvements in speech coder technology since GSM was designed, not to reduce the coding rate but to enhance the voice quality toward fixed-line quality. This coder has already been widely deployed in the US GSM1900 networks and is expected to become widespread in Europe by 1999.

▶ *Combined DECT/GSM handsets:* These are handsets that can work on both technologies, allowing use on DECT indoor systems as well as GSM outdoor systems. At the time of writing, there seemed little support for this feature.

There are other possible evolutionary paths to third generation. The cdmaOne forum has proposed a wideband CDMA system called cdma2000 that is similar to the UMTS proposals, although the bandwidths for the two systems are slightly different because each has been designed to allow easy interoperability with the existing uses of the spectrum in Europe and the United States. The IS-136 system also has an evolutionary path termed D-AMPS++, although this appears to be less well supported than UMTS and cdma2000 and may not become a key contender.

It seems likely that by the year 2005 there will be a new generation of cellular systems capable of higher data rates than can currently be achieved and using dual-mode capability to communicate with satellites when out of the range of terrestrial phone systems.

11.3 Designing the third generation system

One of the prerequisites of the third generation system was that it would be accepted worldwide as a single global system, achieving the design aim of international roaming. However, recently this has started to look like an increasingly unlikely outcome. There are broadly three key players in third generation work—Europe, the United States, and Japan.

Work on designing the third generation mobile radio systems has been difficult. Each country wanted standardization decisions that provided benefits for its manufacturers. The Japanese were keen that the third generation system not be based on GSM because they had not achieved a high penetration of the GSM market. The Americans required a system that aligned with the spectrum spacing they had already assigned for second generation systems from the third generation band. The Europeans wanted a system closely aligned to GSM so that they could maintain their dominant position. The most visible sign of conflict was the selection of the air interface.

Selecting the air interface is only one of a number of decisions. For GSM it has been estimated that the air interface only contained 10% of the specifications; the remaining 90% covered the manner in which the core network (i.e., the switching system and BSCs) interworked. Hence, it might have been thought that it would be the core network that attracted most interest. However, the core network is based on numerous existing standards, such as SS7, which there is little inclination to change;

in any case, changes to the core network do not impact on the spectrum efficiency. As a result, most bodies were prepared to accept that the core network should be based on GSM with the required upgrades.

The air interface decision for UMTS was supposed to be reached in a scientific manner. There were five key proposals for the air interface:

Alpha	wideband CDMA (W-CDMA);
Beta	orthogonal frequency division multiple access (OFDMA);
Gamma	wideband TDMA;
Delta	TDMA with spreading (WB TDMA/CDMA or TD/CDMA);
Epsilon	opportunity driven multiple access (ODMA, a technique whereby mobiles relayed messages to other mobiles).

Groups worked in parallel to evaluate each of the different techniques over a period of approximately one year. Each group was given the same scenarios, and the intention was that the results from each group would be comparable, allowing an access technology to be selected according to which best met the requirements. During this process, manufacturers started to align themselves behind their preferred solution, typically that in which they had the greatest intellectual property rights. The Alpha proposal was based on Japanese suggestions concerning the third generation access technology, which was also similar to proposals from cdmaOne manufacturers for an evolution of their system. Hence, there was some hope that if Europe could accept the Alpha proposal, a unified global standard might result.

At the meeting when a decision was scheduled to be reached about the most appropriate access scheme, the following points were noted.

▶ It was shown that ODMA could be an enhancement to any of the other air interfaces and, hence, could be withdrawn from the decision.

▶ All of the remaining proposals had been shown to meet the UMTS requirements.

▶ All concept groups had performed extensive simulations, but each had tailored the simulations to show their system in the best light, with the result that each group could claim that its system was better than the others!

▶ There was strong support from the manufacturers for the Alpha and Delta proposals that they justified in terms of their technical merits, but was clearly based more on commercial considerations.

▶ The Delta concept made enhancements to its proposal at a late stage, but was not able to perform simulations, so it became even more difficult to assess the relative merits.

▶ There were numerous requirements for the air interface, and it was not possible to reach agreement on which requirement was the most important.

As a result it was not possible to select an air interface scheme on its technical merits, despite the fact that the stated intention was to select the scheme that was technically the most appropriate! This is not surprising given the remarks in Chapter 13; engineers should realize that issues such as the way the standard can be marketed, the management of risk, and the network features provided can be as important as the technical capabilities of the standard. Perhaps it was unsurprising after the TDMA/CDMA debate had characterized the second generation (see Section 14.4) that a technical evaluation was not successful, but it should be remembered that for future standardization any evaluation must look more widely than technical merit.

Eventually, ETSI took a vote on whether the Alpha or Delta schemes was most preferred, but the vote was inconclusive, not achieving the 71% required by ETSI rules to make a decision. At that point the "politicians" suggested that both the Alpha and Delta schemes could be adopted, the Alpha scheme being used for paired spectrum bands and the Delta scheme for the unpaired bands. The engineers were sent to find some justification as to why this would be an appropriate solution. The compromise satisfied both parties who were able to agree on the selection of the air interface. However, at the time of writing it is still unclear whether this will result in a more complex and less efficient radio system than would otherwise have been the case. An article in the *Financial Times* on February 24, 1998, claimed that the reason that a compromise had been reached was because the U.K. government unusually expressed its preference. It selected the Alpha scheme because it hoped it would result in a single global standard. The issue of government intervention is discussed in Appendix B.

Design work on the third generation systems is scheduled to continue until the end of 1999. There will be many more hard decisions en route that will need to be solved in a similar manner to the air interface selection. The result may be a system that the big manufacturers are able to control through their IPR and that is unlikely to upset the status quo in the mobile radio world.

References

[1]   http://www.etsi.fr.

[2]   Zvonar, Z., and P. Jung, eds., *GSM Communications Towards Third Generation Systems*, Berlin: Kluwer, 1998.

Regulators and governments

Increasingly the world of mobile radio communications is being dominated more by the decisions of regulatory bodies than of research laboratories and engineering departments. Radio spectrum, as mentioned earlier, is essential for mobile radio communications; hence, an understanding of the policies adopted in the distribution of radio spectrum is important for the complete wireless professional. Government regulations constrain operators, preventing them from taking certain courses of action and artificially favoring certain technologies by putting in place barriers to others. The other side to regulation is the use of standards, mandated by many European regulators. This section also examines the standards-making process, looking at the way in which the complete wireless professional should partake in standards forums.

CHAPTER 12

Contents

Radio spectrum

I would suggest the taxation of all property equally, whether Church or corporation.

Ulysses S. Grant (1822–1885)

12.1 Introduction

Radio spectrum is the one critical resource for all mobile radio communications. Without access to radio spectrum, there can be no mobile radio. As discussed in Chapter 2, spectrum is a scarce entity and ways are required to ensure that only the most appropriate users are given access to it. Understanding the way in which spectrum is managed and controlled is essential for the complete wireless professional. This chapter looks at spectrum from the viewpoint of the spectrum manager.

12.2 The management of radio spectrum

The world of radio spectrum management is composed of a number of institutions. The most important of these is the ITU. (This was previously known as the CCITT (Central Committee for Telephony and Telegraphy.) The outline structure of the ITU is shown in Figure 12.1. Each sector also has study groups that are not shown.

The International Telegraph Union was set up around 1865 after 20 European states decided to meet to work out a framework agreement for international telegraphy. They also decided on common rules to standardize equipment to facilitate interconnection, adopted uniform operating instructions, and established common international tariff and accounting rules.

Following the invention of the telephone in 1876 and the subsequent expansion of telephony, the Telegraph Union began to draw up international legislation governing telephony. With the invention in 1896 of wireless telegraphy, it was decided to convene a preliminary radio conference in 1903 to study the question of international regulations for radio-telegraph communications. A convention was signed in 1906, and the Annex to this convention formed the first draft of the Radio Regulations.

In 1927, the *Consultative Committee for International Radio* (CCIR) and the *Consultative Committee for International Telegraphy* (CCIT) were established, with the CCIR made responsible for coordinating the technical

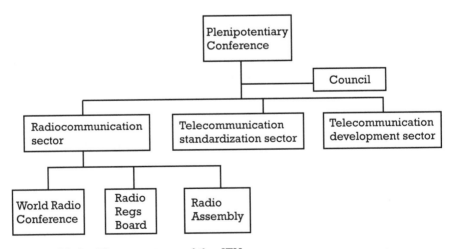

Figure 12.1   The structure of the ITU.

studies, tests, and measurements being carried out in the various fields of telecommunications and for drawing up international standards.

At the 1932 Madrid Conference, the Union decided to combine the International Telegraph Convention of 1865 and the International Radiotelegraph Convention of 1906 to form the International Telecommunication Convention. It also decided to change its name and was known from January 1, 1934, as the International Telecommunications Union. After the first satellite was launched, the CCIR set up a Study Group responsible for studying space radio communication in 1959, and an Extraordinary Administrative Conference for space communications was held in 1963 to allocate frequencies to the various space services. In 1989, a Plenipotentiary Conference decided to increase the importance of technical assistance to the developing countries and set up a *Bureau for the Development of Telecommunications* (BDT) to aid telecommunications development in the Third World.

Each of the key parts of the ITU structure that affect radio communications policy are now considered.

Plenipotentiary Conferences determine key areas of policy and decide on the organization and activities of the ITU. Their decisions are detailed in the International Telecommunication Constitution and Convention. The Plenipotentiary Conference is composed of delegations representing all members and is convened every four years. The duration of Plenipotentiary Conferences is normally limited to four weeks.

The ITU Council is composed of forty-six members of the Union elected by the Plenipotentiary Conference. The role of the Council is to consider broad telecommunication policy issues in order to ensure that the Union's policies and strategy are in accord with the changing telecommunication environment.

Radio communication conferences are held every two years along with a Radio Communications Assembly. The main function of radio communication conferences is to review and revise, as necessary, the Radio Regulations on the basis of an agenda adopted by the ITU Council following consultation of the membership. The general scope of the agenda is established four years in advance and the final agenda is established by the ITU Council two years before the conference. Radio communication conferences may also recommend to the Council items for inclusion in the agenda of a future conference and give its views on forthcoming agendas for conferences.

These conferences are often fraught as each country tries to negotiate a settlement that meets its own interests. Like standardization, discussed in Chapter 13, there is little that is based on engineering judgment. Decisions are made by regulators based on the difficulties they expect in clearing the frequency bands and whether they expect the decision to aid their local industry (e.g., because their industry has a lead in a particular product at a particular frequency). Large organizations often resort to lobbying tactics in order to press their case. Some allocations will be justified on engineering grounds, but typically this justification supports rather than drives the allocation process. At the end of the conference, the ITU recommendations are updated.

Generally, the only engineers able to participate in these conferences are those belonging to the national regulator. These will have studied the problems for many months and understood the potential interference resulting from particular decisions. Engineers from operators and manufacturers can only observe the proceedings.

Radio communications assemblies provide the technical basis for the work of world radio communication conferences; approve the program of work of radio communication study groups; and decide on the priority, urgency, and timescale for the completion of their study. Radio communication study groups are groups of experts provided by administrations and public/private sector bodies. They study technical questions relating to radio communication issues and adopt recommendations. The focus of study is on the use of the radio-frequency spectrum in terrestrial and space radio communications (including the geostationary-satellite orbit), the characteristics and performance of radio systems, the operation of radio stations, and the radio communication aspects of distress and safety matters.

The Radio Regulations Board approves the Rules of Procedure, considers any matter that cannot be resolved through the application of the Rules of Procedure, and performs any duties related to the assignment and utilization of frequencies and to the equitable utilization of the geostationary-satellite orbit. It also investigates cases of harmful interference and formulates recommendations for their resolution.

An ITU recommendation looks something like the stylized version shown in Table 12.1.

Recommendations cover the entire radio frequency band for three different regions of the world, namely:

Table 12.1
A Stylized Example of an ITU Recommendation

Frequency Band 1.455 to 1.460 GHz		
Region 1	Region 2	Region 3
FIXED	FIXED	SATELLITE
aeronautic (note 234)	MOBILE	fixed

1. Europe, Russia, and Africa;

2. North and South America;

3. The rest of the world (China, Australia, Asia-Pacific).

The text in capitals is known as the primary allocation. A user with such an allocation is able to generate as much signal as they like within their allocated band, with no concern over any interference this may cause to other users. Any text in lower case is known as a secondary allocation. A secondary user must not interfere with the primary user, nor must they complain if they receive any interference from the primary user. This only typically works where both users are stationary so that the secondary user can plan their system based on a knowledge of the transmission characteristics of the primary user.

Within Europe, there is a subsidiary spectrum management body known as the CEPT. Members from spectrum administrations in European countries attend meetings in order to decide by consensus on pan-European spectrum allocation issues. Usually, a member country or a standardization body states an interest in a pan-European allocation for a particular application. Study teams are then appointed to assess the spectrum requirements for the new service, the most appropriate frequency bands to use, and whether members can make radio spectrum available for this application. Finally, they recommend what spectrum should be allocated and this recommendation is put to a vote by all the members. In some cases, members may not accept the recommendation due to existing use of the spectrum. If the proposal is generally supported, then the allocation becomes accepted by the CEPT, which may note that the allocation will not become available in some countries in the near future.

Studies into the particular uses of radio spectrum often have little to do with engineering, despite their apparent engineering focus. Typically, they will be staffed by engineers both from the company that wants the assignment and the company that is currently using the spectrum. Because of the difficulties in calculating modern radio spectrum issues, agreement is difficult to reach. An interesting example was when considering the required guard band between TETRA and GSM systems at 900 MHz, apparently a simple engineering problem. The GSM users (who already had a system in operation) wanted a worst case analysis, when a TETRA mobile was only 1m away from a GSM mobile. The TETRA users (who were hoping for the largest possible allocation) wanted an average analysis where the distribution of the interference was determined and some level, such as the 95% probability point, was used to determine the interference. The difference between the two analyses was vast, and both parties were determined that their approach was most appropriate. It is then for the politicians to determine a compromise solution acceptable to both parties.

Most spectrum management work is performed at a national level by a government, or a government-appointed body such as the FCC in the United States. These bodies have control over which particular users are allowed to use the band and police the use of the radio spectrum. They are also responsible for determining the most appropriate manner to distribute the radio spectrum.

The world of spectrum management is becoming increasingly difficult. Pressure on the radio spectrum is growing rapidly from a wide range of sources. There are many new applications that are seeking radio spectrum such as WLL and satellite systems. Existing applications such as cellular radio systems are continually seeking additional spectrum such as the UMTS allocation. Because such users realize the potential profit from the use of radio spectrum, they can often be vociferous in their claims for spectrum, taking recourse to legal action if they are not satisfied. This places pressure on the spectrum manager to be able to clearly justify any allocation and assignment decisions that they might make.

Spectrum managers are also often under pressure from their own governments to show that they are maximizing the value that the country receives from the radio spectrum and to demonstrate that they are acting in a transparent and open fashion. As lead by the United States, governments are also increasingly expecting spectrum managers to sell or lease the spectrum and realize capital gains for the government.

The next section discusses the tools that spectrum managers have available in order to help them overcome these new challenges.

12.3 Modern allocation and assignment methods

In the past, most spectrum management decisions were taken on a judgmental basis. A company would apply for some spectrum and the spectrum manager would assess who was currently using the band and whether the existing user should be displaced to make way for the new user, or encouraged to share with them. Such an approach was successful when the number of applicants was relatively small and the spectrum manager sufficiently skilled, but judgmental approaches are generally not transparent and can be difficult to prove conclusively in a court of law. Increasingly spectrum managers have turned to economic tools to help them with their task of allocating spectrum.

The most obvious approach, and one that gained much publicity in 1997 in the United States, is the use of auctions to distribute radio spectrum. This is intuitively simple—the spectrum that is available is given to the highest bidder. Auctions have much to recommend them, they are relatively simple, they can distribute spectrum rapidly, and they have the happy side effect of raising revenue for the government. They are easily defended in law and can be seen to be transparent and open.

However, there are a few problems with auctions. The first is that radio spectrum must first be cleared of existing uses before it can be auctioned,[1] so the spectrum manager still has to make a decision about which bands to clear for which applications. To some extent, the spectrum manager is helped in this task by international regulations, but these can only provide guidance. The next is that the design of the auction is critically important. Auction theory is a complex area and is explored in more detail in Section 12.4. There have been some auctions of radio spectrum

---

1. Actually, there is a variant of the auction where the spectrum is offered with the existing user (the "incumbent") still in place. The winner must then decide whether to work around the existing user or to compensate them sufficiently that they move to a different frequency band. There are many problems with this approach with the incumbent often holding the new operator to ransom for use of the radio spectrum, and the spectrum manager often ends up intervening to solve the problem.

in the past where, due to poor auction design, the results have been so embarrassing to the government that ministers have resigned. The U.S. auctions have also highlighted another problem known as "winner's curse" where many of the winners have subsequently gone bankrupt. This issue is also looked at in more detail in Section 12.4.

As mentioned previously, auctions can only be used once a decision has been made to clear a band. There are a number of techniques that can be used to help clear a band and to show which users value the band most highly. The most straightforward of these is the use of economic pricing. In this technique, existing users are charged a substantial fee each year for use of the spectrum, equivalent to a rent. This fee encourages them to return any unused spectrum, to make the most efficient use of the spectrum that they already have, and to consider whether moving to a different frequency band might be more economic. To work, the prices need to reflect the demand for the band, such that the most congested bands have the highest prices per megahertz. In principle, the price should be set at the "market rate" for the frequency band, which is the rate that would be paid were the spectrum traded on the open market. The market would push up the price for the spectrum until supply and demand were equal. Herein lies the problem with economic pricing. Determining the market price, without making use of market forces, is very difficult [1, 2] and can only be estimated by looking at the alternatives available to existing users of radio spectrum and the costs that they will incur in taking up these alternatives. The price of the spectrum is then set at a level where it is worthwhile for enough users to move out of the band that there will no longer be congestion.

Spectrum pricing is due to be deployed in the United Kingdom in 1998 after much study, consultation, and legislation; it will be interesting to see how well it works in encouraging more efficient use of the radio spectrum. Like auctions, it also has the happy side effect of raising revenue for the government.

Of course, it is possible to establish a market for radio spectrum and allow real market forces to set prices, overcoming one of the key problems with spectrum pricing. The simplest way to achieve this is to allow users of radio spectrum to buy and sell their spectrum with other users. By this means, a market in spectrum is developed and the prices set by the market can be used to determine the most appropriate use for the spectrum. At present, few spectrum managers are prepared to go as far as trading. They are concerned about the loss of control of the radio spectrum that this would entail because this might result in it being difficult to meet international obligations for the use of the spectrum bands. There may also be difficulties should the spectrum manager wish to reclaim spectrum for any reason—in this case, they would probably need to buy it on the spectrum market. Finally, the revenue from the spectrum will not pass to the government but to the existing users of the spectrum in the form of a windfall gain. To overcome this problem, trading is only normally allowed in spectrum that has first been auctioned so that the value of the spectrum is received by the government in the first instance, although users may then make gains themselves if the value of their spectrum appreciates. Trading has been used in Australia and New Zealand where it appears to be successful. However, both of these countries have the advantage of geographical isolation that allows them to change the use of their radio spectrum with little concern for the interference it might cause to their neighbors.

In order to enable trading, the spectrum manager needs to allow a market to develop. To do so, they need to publish a register of who holds what spectrum and to provide an agency that can put people who want to sell their spectrum in contact with people who want to buy. The agency should also keep track of the sale price of each piece of spectrum as this sends important signal as to the value of the spectrum and the possible need to auction more spectrum in particular areas where there is a high demand.

Finally, the spectrum manager can attempt to assess the value of the spectrum to the country directly. In principle, if the value that the country derives from each of the different uses of the radio spectrum can be estimated, then the spectrum manager can take spectrum from one use that is deriving little value and give it to another use that is deriving a greater value. This approach is known as economic value analysis. To do this, each part of the radio spectrum band is studied to determine what value the country derives from the use. The value is composed of the operators profits and expenditure on the local economy plus the value that the users derive, over and above the fees they pay for using the system. Like spectrum pricing, deriving these values is difficult and experience in the United Kingdom has shown that substantial effort is required to even approximate the value. Nevertheless, the process of attempting the estimation is a valuable one in providing important information to the spectrum manager.

12.4   Implications for the mobile radio operator

Where spectrum managers decide to make use of economic techniques, there are important implications for the operators, both in the manner in which they obtain a license and the manner in which they run their network—for example, the use of technologies providing a high spectrum efficiency becomes even more critical than it was before.

Some users are also caught between spectrum management policies. Market forces are being used to assign their spectrum, but the spectrum manager also wishes to place restrictions on the technology that they can use in their spectrum. To date, there has been a tendency to mandate particular technologies in European countries where the use of GSM is mandated for cellular applications, TETRA for digital PMR, and the *European Radio Messaging Service* (ERMES) for paging. Appendix B discusses whether mandating technologies is appropriate and presents some arguments that can be used if it is decided to attempt to challenge the mandate.

The use of the economic tools discussed in this chapter is almost certain to result in a higher cost to the operator than would have been the case with more traditional spectrum management tools. Many operators have complained that spectrum pricing and auctioning is just another form of taxation that takes money out of the mobile radio industry and increases the cost for consumers. However, the economic evidence is that this is far from the truth. The early mobile radio operators typically made supernormal profits, that is, profits in excess of the returns that could be expected for the level of risk involved. Indeed, some of these early operators made around a 60% return on investment when the sector norm was around 10%. Pricing will reduce these levels toward the norm, as during an auction the rational operator will only bid at a level that still allows them to make an adequate profit. Hence, mobile radio operators in the future can expect to be profitable but not super-profitable, as long as they bid sensibly for the spectrum. Clearly operators who overvalue the spectrum, as some of the PCS operators in the United States appear to have done, run the risk of not being able to operate profitably. The positive side is that the license award is now much more transparent, and rather than spending time and money on lobbying activities often with an unsuccessful outcome, companies now have a clear series of rules that they can follow if they want a license.

Bidding in an auction   Many new operators will face an auction in order to obtain their radio spectrum. It is quite likely that the complete wireless professional will get involved in the process. In order to understand how to devise an auction strategy and how to bid during an auction it is first necessary to understand a little about auction design.

Auctions for radio spectrum are far from simple. Most people initially think of an auction as the familiar bidding at an auctioneer's for some antique, but the auction forms used for spectrum distribution are vastly more complex. Auction design itself is a complex field and there are chairs in auction design within economics departments of some top universities.

In the simplest spectrum auctions, there is only one lot to be sold, for example, a single national license to operate a mobile radio system. A more complex auction might offer say two or three identical national licenses and there might be a rule that each bidder can only obtain one license. The next stage of complexity is that the licenses may not be identical, for example, in different frequency bands, and also where bidders are allowed to obtain more than one lot. The final level of complexity is where there are multiple lots both in different frequency bands and different geographical areas. Bidders might want to obtain two adjacent frequency bands throughout the southern part of the country and may have to win, say, 20 lots to achieve this. The last form is that closest to the PCS auctions in the United States. Some of the different auction types that can be used to overcome these problems are as follows:

▶ *English auction:* In an English auction, a single item is offered and bidders increase their offers until nobody wishes to bid more highly than the last offer. It is important to understand with this auction that the winner has not paid their own valuation of the item but instead has paid slightly more than the second highest bidder's evaluation of the value of the item. This auction form is well known but generally judged inappropriate for spectrum where sealed bids are preferred. Without modification it cannot cope well with multiple related lots.

▶ *Sealed bid first-price auction:* In this auction, bidders submit a sealed bid and the highest bidder pays what they bid. Bidders, however, will not bid at their valuation of the lot but at a price that they estimate is higher than what the other bidders will bid. If this estimate is higher than what they are prepared to pay, then they will not bid.

Complexities then arise since all bidders are trying to second-guess the other bidders. The result is often more of a lottery than other auction methods.

▶ *Second-price sealed bid auction:* In this auction, bidders submit a sealed bid. The highest bidder pays the price offered by the second highest bidder. This overcomes the problems of the sealed bid first price since bidders now bid their full valuation. The problem with this form of auction is that the public often perceives that the government is "giving away" the spectrum for less than the bidders were prepared to pay for it. This, of course, is not true, since the bidders would not bid their full valuation if they knew they had to pay it, but it is difficult to convince the public of the correctness of the auction. In New Zealand, public anger about the situation where there were only two bidders—the first bid many millions, the second only $1, and the first bidder was awarded the license for $1—resulted in a ministerial resignation. Hence, this form of auction is now rarely used.

There are a number of multiple object auctions that are complex and only one has been used for radio spectrum. The single key type for radio spectrum is the simultaneous multiple English auction. It was devised by the FCC in 1994 and has been revised and updated in the light of a number of auctions. For every lot to be sold, there is an unlimited number of rounds using sealed bids. At the end of each round, the highest bid for each lot is made public. For the next round, each bid must be above a minimum increment over the previous bid price. All the lots remain open until there is no bidding on any lot during a particular round. The advantages of this form of auction are:

▶ Bidders who want a number of related lots can make sure that they obtain them since the bidding remains open until all lots have concluded or can withdraw early when they realize that the bidding will rise higher than their valuation (there is some risk that they will have bid the highest price on one lot before realizing that the other lots are too expensive and then have to withdraw from their "winning" lot, incurring a financial penalty for doing so).

▶ Bidders gain information about the value others place on the license during the bidding process and can adjust their valuation if it appears out of line with that of others (e.g., if others appear to be bidding much less, the bidder might conclude that they have been optimistic about the risks).

▶ Experience has shown that these auctions are effective in selling similar lots for similar prices, allowing bidders to accumulate the lots they require.

The full set of rules is necessarily very complex and includes rules about timing, minimum increments, deposits, payments, and defaults, for example. This complete rule set has been found to be necessary to prevent bidders trying to find means to "cheat" the auction process.

Preparation for this form of auction by the bidder starts with their valuation for the license. Valuation of the radio spectrum is a complicated process but one that each bidder must carefully perform. Essentially, the bidder builds a business case as explained in Section 8.2. The bidder then decides what return on capital they require and any extra return they can find is their valuation of the radio spectrum. Because building a business case relies on difficult assumptions about subscriber numbers and tariff predictions, it is not surprising that different prospective operators will come up with different valuations. The basic bid strategy is then simple—continue raising your bids on each lot in which you are interested until the valuation price is reached. However, if some lots are shaping up more cheaply than others, then it may be appropriate to modify bidding strategy in order to focus on a different geographical area or a different frequency band. Actually bidding is more complex because of the need for a computer to monitor all the different lots. Further, because the duration of each round may be quite short (perhaps half an hour), it is necessary to preplan different strategies, looking at a range of possible scenarios. Game theory and "war game" preparation is a useful tactic here. Most bidders will run bidding software on their computers that knows all the myriad of rules so that they do not infringe upon any rules and can place bids in a timely fashion. Bidding will normally be via the Internet using computers to prevent all bidders having to be in the same building.

The U.S. PCS auctions have cast some doubt on the efficacy of this auction form because many of the highest bidders have defaulted on their bids and returned the spectrum to the government. This has caused many to suggest that the auction form is flawed since it is failing to distribute licenses quickly. This problem with auctions is well known and is called the "winner's curse." Essentially, the winning bidder is the one who most overvalues the market. When they come to implement their network, their overvaluation becomes apparent and they may go bankrupt. This is not problematic if the spectrum can be rapidly reclaimed and reauctioned but is perceived by the public as a failure of the bid process. There are a few rules that can be established to overcome the winner's curse:

▶ Require bidders to demonstrate that they already have funding to finance their bid before they can make the bid.

▶ Only accept bids from large companies with a good "track record" in running networks (but this tends to reduce entrepreneurship).

▶ Provide information as to the value of the licenses in similar situations (perhaps in other countries).

These will tend to reduce the dynamism in the bidders, and hence in the manner in which networks are run, but will also reduce the risk of defaulting. It is up to the governments to set an appropriate level of risk.

12.5 Government policy

Governments play a key role in the area of mobile and fixed communications. Most of the government regulators are generally concerned with fixed operators [3]. Typically, they try to ensure that the incumbent fixed operator, the PTO, does not abuse its monopoly position and indulge in unfair competition. Fixed operators are generally outside the scope of this book (except where they are also mobile operators), but a good description of the role of regulation on fixed operators can be found in [4].

There are a number of regulatory matters that impinge upon mobile radio operation.

Coverage requirements   Governments are typically keen that giving out a mobile radio license results in the provision of mobile radio services to most of the population. If, instead, a mobile radio operator decided to provide a service only in the capital city, which might be a quite lucrative mode of operation, the government would be concerned that mobile communications had not been provided to the whole community and as a result the true economic benefit of the use of mobile communications would not be realized. In this instance, the government would like a means to revoke the license and to give it to a different operator prepared to provide better coverage. The control that the government uses over the mobile radio operator is typically a coverage requirement. The mobile radio license will normally say that a certain percentage of the population (often 90%) is covered within a given timespan (often 5 years). If this coverage is not met, then the government can withdraw the license.

There is some debate as to whether coverage requirements are actually a useful tool for the government to adopt. To date, all mobile operators have surpassed their coverage requirements long before the date in their contract as they have realized that good coverage is essential in winning new customers. Further, if the operator decided that meeting the coverage obligation would actually result in an unprofitable network, they might be inclined to bid less for the license or not at all, reducing the overall value of mobile communications. In a free market where there is competition, it would seem that coverage obligations are not appropriate. Finally, it is far from clear whether such obligations can actually be enforced in any case. Say an operator had 80% coverage, rather than 90% (and measuring coverage to this accuracy is in itself a problem), then it is unlikely that the government would withdraw the license since such a move would result in many users being unable to use their mobile phones, which would be unpopular.

Whatever the efficacy of coverage obligations, it seems likely that they will be a part of mobile radio licenses for some time.

Fixed/mobile integration   For some time regulators have differentiated between fixed and mobile licenses. In particular, some fixed operators have been prohibited from holding mobile licenses in order to increase competition. In some countries, the converse is also true, that mobile operators have been prohibited from offering fixed services. However, the distinction between fixed and mobile communications is becoming increasingly slim. Fixed operators are now placing DECT base stations at the end of fixed lines and claiming that this is just an extension of the fixed service. Mobile operators are using their networks to provide WLL connections, offering a fixed service. It seems likely that in the future the regulator will increasingly be unable to draw a distinction between fixed and mobile services and will stop attempting to make companies provide only one or the other.

Number of mobile operators   One of the key decisions that a regulator needs to make is the number of mobile operators that should be allowed. Coupled with this is a decision as to how much radio spectrum each operator should be given. Although the limits of the radio spectrum place a maximum on the number of operators, this number is often further restricted for competition purposes.

Selection of the number of operators is a very difficult problem. On the one hand, if there are too many operators, too much will be invested in infrastructure and communications will be expensive for the population. On the other hand, if there are too few operators, then there may be insufficient competition to drive prices down. This is a massively complex situation in which engineers can do little to calculate the most appropriate outcome. Experience has shown that two operators are probably insufficient since a "cozy duopoly" tends to form. At present in most countries where five have been tried, two have merged, resulting in four, so the general assumption is that there should not be more than four. The difficult question is then whether there should be three or four.

The complexity and importance of the situation can be seen by looking at what recently happened in a European country. This country already had two operators at 900 MHz and was forming a new law to allow additional operators at 1,800 MHz. The initial position, apparently formed without too much thought, was that there would be one additional operator at 1,800 MHz. This was borne out by the fact that there was thought to be only sufficient radio spectrum for a single operator at 1,800 MHz, although more was becoming available later. However, when the legislation to allow a single operator was placed before the government, some objections were raised that more than one operator should be licensed. In particular, EC law suggests that there are very few cases that one operator should be licensed, with a shortage of spectrum being one of the few allowable reasons.

At this point a team of technical and economic experts were called in. The technical analysis attempted to establish how much spectrum would be required. However, as shown in Chapter 2, a mobile radio network can be built with limited spectrum if there are a sufficient number of cells.[2] For the operator, there is a tradeoff between cost and spectrum. Hence, the technical specialist rapidly concluded that from a technical viewpoint

---

2. The minimum amount of spectrum is equal to the carrier bandwidth times the cluster size. In the case of GSM, with a cluster size of 21 for the control carriers, this equates to 2×4.2 MHz.

there was actually sufficient spectrum for *three* operators; however, they might not be able to operate economically due to the number of cell sites required. The economic experts studied the situation and suggested that appropriate auction methods could be devised to let the market decide whether there should be one or two operators. Essentially, a form of auction could be used where two licenses were initially offered and, if there were not two appropriate bids for these licenses, the auction would be rerun for a single license. The economists stated that this was the best way to proceed since only the market had sufficient knowledge to judge how many operators were required.

The government decided to ignore this advice and pressed ahead with the legislation for a single operator. For a time this appeared to be succeeding, but in the final stages of legislation the legality of the move was again questioned and it was finally judged to be inappropriate. At this point, the draft legislation was withdrawn and replaced with legislation that allowed two operators. Rather than just demonstrating a victory for consultants over governments, this shows the regulatory difficulties in determining how many mobile licenses should be awarded.

Distribution channels   Another area where regulators can affect mobile radio policy is in the manner in which operators are allowed to sell their airtime to the public. In some countries the regulator has insisted that operators should not be allowed to sell directly to the public but instead should make use of service providers—third parties buying airtime wholesale from the mobile operator and selling it through a chain of stores to their customers. This clearly has an implication on the manner in which the mobile operator is structured but has limited implications for the engineering staff.

References

[1]   NERA, Smith, *Review and Update of 1995 Economic Impact Study*, available from the U.K. Radiocommunications Agency, 1997.

[2]   NERA, Smith, *Study into the Use of Spectrum Pricing*, published by the U.K. Radiocommunications Agency, Apr. 1996; also available from http://www.open.gov.uk/radiocom.

[3]   Calhoun, G., *Wireless Access and the Local Telephone Network*, Norwood, MA: Artech House, 1992.

[4]   Clarke, M., *Networks and Telecommunications: Design and Operation*, New York: John Wiley, 1997.

CHAPTER

13

Contents

Standardization

The minute you read something you can't understand you can almost be sure it was drawn up by a lawyer.

Will Rogers

13.1 Introduction

Standardization is becoming increasingly important in the world of mobile radio. In cellular, PMR, paging, and cordless telephony, standardized products are starting to dominate the world market. With standardization of third generation products underway it seems likely that the domination of standards will continue into the foreseeable future. The complete wireless professional may become involved in standards in a number of different ways:

▶ Directly, as part of a team responsible for writing a new standard or making amendments to an existing standard;

▶ Indirectly, as the ongoing standardization affects product strategy;

▶ As an end user of the standard, either in an operator or a manufacturer, trying to understand a published standard and develop an appropriate offering based on the standard.

In all these cases it is helpful to understand the role of standardization bodies, the manner in which mobile radio standards are written, and the choices made during standardization. This chapter describes the standardization process in general and presents a case study of a particular standardization exercise to illustrate some of the key issues.

13.2 Standards-making bodies

Standards-making activities occur in three key places around the world: Europe, the United States, and Japan. In Europe, standards are developed by the ETSI; in the United States, in the main by the TIA; and in Japan by the *Association of Radio Industries and Businesses* (ARIB). The most successful of these bodies have been ETSI and the TIA, and this section will focus on these two.

ETSI ETSI is based in France but is responsible, under the auspices of the EC and CEPT, for the development for all mobile radio standards for Europe. A simplified diagram of the structure of ETSI is shown in Figure 13.1. As can be seen, ETSI covers the standardization of all telecommunications equipment including fixed and mobile equipment. Each different area tends to have its own *technical committee* (TC) that, when a particular standard is being developed, tends to be given its own working team of individuals who work full time at ETSI on the standard for the few years that its development takes. The key cellular standardization takes place within the *special mobile group* (SMG). This does not follow the same naming convention as the rest of the development groups since SMG was originally formed within the CEPT and was then "transferred" across to ETSI during the development process of GSM (which takes its original name of Group Special Mobile from the committee performing its standardization).

Within SMG there are a number of subgroups that change on a regular basis, but the key subgroups are:

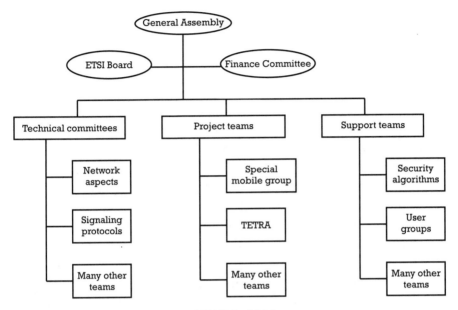

Figure 13.1 The structure of ETSI in 1998.

▶ *SMG1—User requirements:* This group captures the user's needs and sets them out in clear documents known as requirement documents.

▶ *SMG2—Radio aspects:* This group studies issues associated with radio transmission and writes the associated standards.

▶ *SMG3—Network aspects:* This group designs the protocols required for correct operation of the network.

▶ *SMG4—Data systems:* This group considers issues particular to data transmission.

▶ Numerous other groups considering aspects such as SIM cards and testing.

Work within ETSI tends to start when the EC notes the requirement for a new standard. The EC will request ETSI to identify the required resources and set up a *project team* (PT) for the duration of the standardization process. The project team is staffed by engineers on secondment from industry, paid for by the EC, for the duration of the standardization. For the major standards the funded work can run to thousands of man-days of paid activities costing many millions of pounds. Industry is then invited to partake in the standards process. Only European companies can belong to ETSI, although U.S. companies such as Motorola qualify as a result of their European activities. Typically, monthly meetings are held by each group at which all interested parties attend, documents are presented, and key decisions are taken by consensus, or by voting if required. This process is described in more detail in the next section.

ETSI standards are published and are available to ETSI members from the ETSI document center. Once they have been officially approved by ETSI, they typically become mandatory for that particular technology within the EC (see Appendix B for more details of mandating standards). More details about ETSI can be obtained from [1].

U.S. standards bodies There are two bodies in the United States involved in standardization. Most mobile standardization is performed by the TIA; however, the *Alliance for Telecommunications Industry Solutions* (ATIS) also plays a role. Historically, standards making in the United States was performed by AT&T, then the only organization with an interest in telecommunications standards. As the industry was liberalized, trade bodies became more important. The TIA took over most of the standardization of wireless systems, whereas ATIS looked after wireline standardization. However, during the PCS auctions, the T1 committee of ATIS took responsibility for standardization of the GSM1900 solution within a group known as T1P1. This group performed the minor modifications on the GSM standard required for North American operation. Nevertheless, the bulk of the mobile standardization work is performed within the TIA, which is now examined in more detail.

The TIA has five product-oriented divisions, namely:

- User premises equipment;
- Network equipment;
- Wireless communications;
- Fiber optics;
- Satellite communications.

These address the legislative and regulatory concerns of the product manufacturers and prepare standards dealing with performance testing and compatibility. TIA is accredited by the *American National Standards Institute* (ANSI). It tends to firstly give standards *interim status* (IS) and then, after appropriate testing, full ANSI accreditation.

Within the wireless communications division there are three subdivisions:

1. TR-8, which develops mobile and personal private radio standards for PMR usage;

2. TR-45, which develops public mobile and personal communications systems standards;

3. TR-46, which develops mobile and personal communications standards for the 1,900-MHz band.

In practice, TR-46 now performs little work, with the key cellular standardization being performed in TR-45. TR-45 is subdivided as follows:

▶ TR-45.1 deals with the analog standards AMPS and N-AMPS.

▶ TR-45.2 deals with the intersystem interface know as IS-41 (the equivalent of the MAP interface within GSM).

▶ TR-45.3 deals with D-AMPS, also know as IS-136.

▶ TR-45.4 deals with the "A" interface that connects base stations to switches.

▶ TR-45.5 deals with cdmaOne.

▶ TR-45.6 deals with the emerging cellular packet data system.

U.S. standardization is somewhat different from that in Europe. The standards bodies have no permanent staff who can be assigned to a project team to progress a particular standard—all members are part-time attendees from their companies. As a result, standardization tends to be slower and it is often more difficult to achieve consensus than within ETSI. In practice, most standards get developed in a quite different manner. A company with a good idea will persuade a number of other companies to form an ad-hoc group that sits outside any standards bodies. This group will develop a specification and design that they will then present to the TIA in almost finished form. If there is sufficient weight behind the proposal, it will be accepted and a new division within TR-45 will be established to develop the proposal. The determination of "sufficient weight" is an inexact science that depends more on politics than on quantitative measures of support. This use of ad-hoc bodies means that there can often be more than one standard for the same area, such as both the D-AMPS and cdmaOne standards for digital cellular. Such an approach would not be allowed within ETSI, where one of the objectives is a single unifying standard.

Because of the major differences in the manner in which standards are developed, the European and U.S. ways have quite different characteristics. The European way has the advantage that any standards produced are typically highly tested, unlikely to have significant errors, and generally carefully designed to meet the perceived requirements of the users. In contrast, U.S. standards may be developed without consideration of the user's needs and may achieve standardization before comprehensive testing. However, the European way allows only one standard to be designed in one particular manner. Contributors who disagree with the standardization work can make objections, but if these are overruled by the committee there is little that they can do. The U.S. method allows those unhappy with one particular standardization route, such as Qualcomm, to embark on a parallel route in a manner that appears to some to be more democratic. In general, Europe is providing the world's most popular standards, generally due to their careful design and their support by the majority of the world's leading manufacturers. However, without the U.S. approach to standardization, it is unlikely that there would be a CDMA-based cellular standards available to date.

More information about the TIA can be found at [2] and about the T1 committee of ATIS at [3].

13.3 Writing standards

The writing of standards is quite different from almost any other engineering discipline and something that takes some getting used to. An ETSI standard is normally divided into three stages:

1. The first stage is a requirements specification that sets out the requirements on the item being standardized.

2. The second stage is a logical description of the manner in which the item will be designed without being specific to any particular realization—for example, it might suggest that group calls should be achieved by having one shared channel in each cell.

3. The third stage is the production of the final standards document; this stage normally lasts much longer than the first two stages.

In principle, each of these stages should be completed before the next stage starts. In practice, the first stage is typically only partially complete before the second stage starts. As implementation problems are discovered in the second stage, these may be fed back to the first stage committee who may agree to alter the requirements if they are likely to be unduly expensive to provide. The same developments might happen between the second and third stages. Only as all the stages are coming close to completion are the early stages likely to be frozen.

The process of writing documents by committee is an interesting one. Typically there will be a special committee established as an offshoot from one of the permanent committees that will be responsible for developing the standard. This committee will have three parts corresponding to each of the three phases of standardization. Each of these parts will meet, probably on a monthly basis. At the meeting, the first activity is to number and organize the input documents. Document control is essential in standardization bodies where otherwise matters can quickly get out of hand. The documents are then assigned to agenda items. As the meeting progresses through the agenda, each person who has provided an input document is asked to present his or her document. This typically means summarizing the document verbally and drawing the attention on the delegates to the key points in the document. There will then follow a period of discussion on the document during which the other delegates will attempt to find problems with the document. The author will take note of these comments and revise the document accordingly, either for input on the next day of the meeting or at the next meeting.

A volunteer is normally appointed to be editor of the specification. This person is responsible for keeping the latest version of the specification, for presenting it at each meeting, and for revising it in line with comments. Others at the meeting may provide additional text for the standard in their input documents and, if it is agreed upon, will pass this text to the editor for inclusion in the specification. There are strict editorial guide-lines for the format of specifications and the manner in which revisions to the specification are made.

This process works well while there is always someone who is willing to provide an input document in all the areas where input is required and no two people are working on the same input document. In practice, the problem of nobody providing a document on a particular area is generally solved as delegates become aware that a lack of document will slow the standardization and eventually agree to take on the work to prevent deadlines slipping. The problem of more than one input document on the same area is more difficult to solve, but one that happens frequently. Dif-ferent individuals will often have different ideas on the way that a par-ticular problem is best solved. Both will produce input documents with different solutions that they will present. The rest of the committee will assess the different documents and, through their comments, will indi-cate their preference for one or the other. Because it is a standards com-mittee, a decision has to be made democratically, and in some cases a formal vote will be taken on which of the different inputs should be used. In practice, matters rarely come to a vote. Those writing the input docu-ments will realize from the comments and from the discussion in coffee breaks whether their ideas find much favor and will tend to withdraw unpopular suggestions before a vote is reached. Seemingly intractable arguments do arise from time to time, and the normal approach is to then refer the problem to the permanent committee who can judge the most appropriate solution.

The advantages of such a process are that it is truly open, democratic, and the process of peer review and criticism generally results in a standard with fewer errors than might otherwise have been the case. The disad-vantage is that standardization in this manner is very slow, with work only really progressing once a month and in a costly manner to the parties involved who continually send key engineers to meetings all across Europe. Matters are sometimes made worse when large manufacturers send new engineers to ETSI meetings as part of their training, particularly on the GSM standard. Time is then required in the meetings to explain some of the more complex points of the standard to these new delegates. This can occur at almost every meeting.

When actually authoring a particular input document, the author needs to use very precise and terse language. If the text is to form part of the final standard, it must be completely clear and unambiguous. For example, in English the words *should, might, may,* and *could* are all ambiguous—it is not clear whether a particular action is required or not required. ETSI guidelines mandate the use of *shall* to indicate that a particular action must be performed and *may* to indicate that performance of a particular action is optional. For the stage-one standards, the text may read like a normal document. By the stage-two standard it is already becoming terse but might, in some cases, offer some explanation as to why certain options have been selected. By stage three, the specification reads a little like a computer program with no comments, showing only the actions that the mobile and the network must take in a range of different situations.

As a result, reading specifications is extremely hard work. To understand the functioning of GSM purely from reading the specifications would be nearly impossible, which is why there is a thriving industry writing books that effectively decode the specification documents.

References

[1]   ETSI website is at http://www.etsi.fr.

[2]   TIA website is at www.tiaonline.org.

[3]   T1 committee of ATIS website is at www.t1.org.

Becoming a better wireless professional

A complete wireless professional needs to be a well-rounded individual. In addition to having a good understanding of the underlying concepts, the complete wireless professional will need to handle issues of conflict in a competent and professional manner, have a good understanding of management, and conduct his or her career in a manner that will enhance his or her ability to provide good engineering advice. This part looks at how to become a better wireless professional.

CHAPTER

14

Contents

Areas of conflict

The most savage controversies are those about matters as to which there is no good evidence either way.

Bertrand Russell

14.1 Introduction

Conflict has been a consistent feature of the world of mobile radio in recent years. A wireless professional is almost certain to become involved in some form of conflict in his or her career, and the better ones will frequently tend to be involved as they get brought into the front line as "elite troops."

Conflicting situations are almost always due to different manufacturers trying to establish their preferred technology in order to sell more product and increase profitability. What almost invariably happens is that the two warring parties both claim that their product is the best, often for the same reasons. When mobile radio systems were simpler, it might have been possible to resolve these conflicts in a straight-forward manner, but mobile radio systems have become sufficiently complicated that typically the conflicts cannot be solved by a rigorous mathematical analysis or simulation. The net result is that both sides can continue to make claims, supported by their own, favorable, analysis, which independent arbitrators (of whom there are very few) are unable to disprove.

These conflicts have become increasingly intractable and the engineers involved in them have tended not to behave in a manner most fitting to their profession. The complete wireless professional should understand how to behave in a situation of conflict and how to act in such a manner that the best interests of all concerned can be realized. It is important to remember the following points:

- In an argument where certain technical parameters cannot be calculated accurately, it is often better to say this than to use approximations that cannot be universally accepted.

- Bitter propaganda battles seem to do neither side any favors. Using different tactics may be more appropriate.

- It is actually quite rare for the best technical solution to be the that which is accepted. A good example of this is the VHS versus Betamax battle. Often, other economic and market-related factors are more important than engineering issues.

- Many conflicts are not actually about technical matters, however it may seem, but are about patents, intellectual property rights, politics, and market position.

To illustrate some of the issues involved, a few of the most prominent conflicts in recent years are considered in more detail here. What will become apparent is that it is rarely possible to provide a clear-cut solution to the conflict but that careful analysis can reveal the key issues. The conflicts considered are:

- TETRA versus GSM;

- PHS versus DECT;

- CDMA versus TDMA.

14.2 TETRA versus GSM

14.2.1 Background to the debate

As discussed in Chapter 4, TETRA is the new European standard for advanced PMR applications. During its design it was intended to be used by all major PMR users and by many of the smaller scale users. Three manufacturers in particular—Motorola, Nokia, and Philips (now SiMoCo)—invested substantial effort in developing the standard and in product design. However, during the standardization work, the International Railway body, the UIC, made the decision that its next mobile radio system would be based on GSM rather than TETRA and initiated the standardization work to add PMR features to the GSM standard. As a result of this, other PMR users were faced with a choice as to whether to adopt TETRA or GSM. TETRA manufacturers were horrified by the possibility that their marketplace might be suddenly reduced and campaigned vociferously to demonstrate why GSM could not possibly meet the needs of PMR users. Meanwhile, the UIC and some of the GSM manufacturers were keen that other PMR users adopt GSM in order to spread the cost of the necessary modifications over a number of other users. The result was conflict.

14.2.2 Evaluation of the technologies

The TETRA standard was conceived to provide a European-wide standard digital PMR technology, with the associated benefits of open sourcing, economies of scale, and roaming possibilities. GSM was conceived to provide public cellular communications. However, the European railways decided to select GSM rather than TETRA as their standard because at the time there was a lower risk associated with GSM, it met their timescales, and GSM's enhanced roaming capabilities compared to TETRA were judged to be an important feature for high-speed international trains.

In addition, there are a number of proprietary digital PMR products from companies such as Ericsson, Matra, and Motorola and a U.S. standard called APCO25, discussed in Chapter 4. These could offer an alternative to TETRA or GSM but do not provide the multisourcing capabilities that generally make open standards so attractive. Further, government bodies will be under some pressure under EC procurement law to procure an open standard for their radio system that may remove the choice of proprietary systems for some users in this area. For these reasons proprietary systems will not be considered further here.

When looking at the functions of GSM and TETRA, what is apparent is that both standards largely meet the major needs of all user groups. Both will support a wide range of voice and data capabilities. TETRA has a number of additional features over GSM-ASCI, which are:

- Back-to-back operation and repeater operations with one mobile inside a coverage area able to relay the signal to a mobile outside the coverage area; direct-mode facilities are not envisaged within GSM, with dual-mode mobiles providing the only feasible direct-mode option;

- Ambient listening (the ability for a controller to activate a mobile without intervention from the user so the controller can listen in the case of hijacks and similar incidents);

- Pseudo-open channel where a user group can withdraw certain specific radio channels from the pool of channels available to all users to provide a minimum guaranteed level of service to important users;

- Load apportionment so that in overloaded situations each user group gets a fair share of the resources;

- Barring of direct access to certain types of calls and so insisting they are routed through the dispatcher;

- Dynamic group allocation where groups can be created by a dispatcher very quickly for a particular incident—the ASCI features only allow fixed group allocation.

In essence, GSM is better at wide area coverage, roaming, international operations, billing, and network management—everything one might expect from a cellular system. TETRA is better at flexible deployment in both small and large configurations and a range of services mainly targeted at the emergency services.

There is some risk that TETRA will not have all the facilities promised when it first arrives; the number of data services was recently dramatically cut during the standardization process, and some manufacturers are seeking reduced functionality TETRA equipment in order to minimize their risk, while the number of GSM services is ever-increasing. Equally, there is some risk that the GSM ASCI features will not be widely implemented.

TETRA requires a smaller bandwidth than GSM for voice communications. TETRA systems can provide 4 voice channels in 25 kHz (6.25 kHz per channel) whereas GSM currently provides 8 voice channels in 200 kHz (25 kHz per channel) and will change to 16 voice channels in 200 kHz (12.5 kHz per channel) with the imminent introduction of the half-rate codec.[1] However, it is important to consider the relative cluster size of the two systems. The GSM specifications show that it requires a *carrier-to-interference* (*C/I*) ratio of 9 dB, whereas the TETRA specifications show a *C/I* ratio of 18 dB. According to Lee [1] the cluster size is given by

$$K = \sqrt{\frac{2}{3} \cdot SIR} \qquad (14.1)$$

Hence, it can be calculated that the (theoretical) minimum cluster sizes for GSM and TETRA are 2.3 and 6.5, respectively. Taking these factors into account, the full-rate GSM coder provides up to 17 voice channels per cell per megahertz, the half-rate GSM coder provides 34 voice channels per cell per megahertz, and TETRA provides 24 voice channels per cell per megahertz. Given that most GSM operators tend to deploy the full-rate coder, TETRA provides a gain of around 40% in spectrum efficiency. This improvement in bandwidth efficiency offered by TETRA could be extremely important for users who are in a spectrum-limited situation.

Despite the additional functionality offered by TETRA, for many users cost will be the overriding consideration. Economic issues are considered in more detail in the next section.

14.2.3 Economic comparison

Once TETRA is available and well established to the extent that analog PMR use is rapidly decreasing, a PMR user might face the following alternatives:

---

1. Although most public GSM operators are not planning to deploy the half-rate coder because of its perceived low voice quality, voice quality is less important to PMR users who may prefer the high-spectrum efficiency that results.

1. Self-provide its own PMR system based on TETRA (or potentially a proprietary technology);

2. Move to a shared (PAMR) system based on TETRA;

3. Self-provide its own GSM system;

4. Move to a public cellular GSM system.

The basic problem with GSM for private users is that it was not designed for small-scale deployment. Recently manufacturers have announced systems designed for only 10,000 users, which represents a reduction over previous figures of 100,000 users but is still excessive for most private users. Each cell provides a minimum of 7 voice channels (15 with the half-rate codec) and the smallest deployment requires about 2×4.2 MHz of spectrum.[2] Unless this sort of capacity is required, GSM will be extremely uncompetitively priced for the private user. Only organizations such as the railways and the police could realistically consider operating their own GSM system.

When comparing the choice between self-ownership of TETRA and GSM (options 1 and 3, respectively) a number of issues are relevant. Cost is clearly important, and the relative costs of the two systems depends on the coverage and capacity required, with GSM generally being favored where high levels of capacity (typically above 4 channels per cell) are necessary. Functionality is also important and some user groups will select one of the two systems because it offers the particular facilities they require. In this area there have been conflicting choices with the railways selecting GSM, the U.K. Home Office indicating a preference for TETRA, the French Police choosing a proprietary system called TETRAPOL built by Matra, and the Basque police selecting GSM!

For most users, self-provision of a GSM system (option 3) is not a feasible possibility. Of the remaining three options, the use of shared PAMR or cellular systems (options 2 and 4) depends first on the existence of appropriate networks and then heavily on the tariffs set by the operators.

Regarding the potential of shared TETRA system, a number of national PAMR operators have signed the TETRA *memorandum of understanding*

---

2. Based on one carrier per cell with a 21-cell repeat pattern for the control channels.

(MoU) suggesting they have some interest in this area. In addition, there are the changes likely to be brought about by the *private finance initiative* (PFI) in the United Kingdom and similar initiatives in other countries. In essence, this seeks to encourage the private sector to provide infrastructure and supply services to the public sector. This is the basis on which the U.K. Home Office is currently working with its *public safety radio communications project* (PSRCP). The result of such initiatives is that users who historically have owned and operated their own radio system are considering sharing systems with others. Either of these may lead to a shared TETRA system.

Concerning the GSM operators, some are clearly interested in attracting PMR users onto their network, first through a modified tariff and then at a later stage through the addition of the required functionality. While the provision of appropriate tariffs without any additional features may be attractive to some PMR users, current GSM networks lack many of the features that many PMR users would find essential—in outline, the ASCI features. Without these features, the attractiveness of any GSM offering would be significantly reduced.

If we assume that shared TETRA networks and attractive PMR offerings from cellular operators do emerge, then the smaller users will face a three-way choice between options 1, 2, and 4. In the instance where they only require coverage of a small area such as a single site but generate a great deal of traffic within that area, self-provision using TETRA is likely to be the most attractive alternative (option 1). Because of the likely cost increase of TETRA over current analog systems, the threshold traffic level where self-provision is economic is likely to rise from the present level with the result that some current PMR users will find it more economical to move to a shared system.

If coverage over a large area is required, then shared systems are likely to be more appropriate. Whether this is TETRA or GSM will depend on cost and functionality. It is too early to say at this stage what the relative costs will be. While the GSM cellular operators do not need to build a network from scratch, they will be restricted in the tariffs they offer if they do not wish to cause migration of existing users. Early estimates are that GSM systems may offer lower cost but also lower functionality. This should become more apparent in the next few of years.

For the larger users, the choice between shared and self-provided systems depends upon many factors. In the case of government bodies, the

PSRCP is likely to force a move to shared networks. In the case of bodies such as the railways, estimations of traffic levels will allow the comparison of tariffed costs against the cost of ownership to be performed.

14.2.4 Analyzing the debate

Looking at the facts presented in the preceding analysis and assuming that these provide a fair representation of the situation, it would appear that the key issues and uncertainties are the following:

▶ TETRA will apparently provide a greater functionality than GSM, but it is not clear to what extent this functionality will be valued by users. Certainly some users will value it more than others.

▶ TETRA may provide a greater spectrum efficiency than GSM, but this remains to be proven and it is not clear to what extent this is an issue for the potential users.

▶ TETRA will probably be more expensive than GSM for most deployments, but there will remain situations where the converse is true.

▶ The optimum choice for both systems depends on manufacturers providing the technology with all the features present and operators deploying the technology.

▶ There are many confusing issues and unknowns such as the new funding process in use by some public bodies that may further cloud the issue.

How is the complete wireless professional to make sense of all these arguments? Many of those involved in the debate could see little further than the argument about functionality, pointing out that since TETRA had better functionality it was the best choice. However, decisions as to which radio system to adopt are now generally made on an economic basis, looking at whether a more expensive system will provide benefits sufficient to offset its increased price. With further analysis, it became clear that for most of the users, the additional features offered by TETRA did not have much value attached to them and hence the argument about additional functionality was false. The key exception to this was the police who, it appeared, did value the additional functionality sufficiently to pay more for the technology.

The complete wireless professional would have noted by this point that, in fact, the engineering issues are not the key points at stake here. Also, he or she will have noticed that many of the issues depend on the situation—for example, airports are in a quite different situation to taxis and, hence, it is not possible to provide any general answers to the debate. Further, the outcome is dependent on the strategy of manufacturers and operators. Typically, it is not possible to predict the strategy that manufacturers and operators will adopt, and long-held strategies can be changed overnight. For example, if no cellular operator announced that they were to offer the ASCI features, then the debate as to whether TETRA or GSM was more appropriate would change dramatically overnight.

The complete wireless professional would be very wary of taking sides in this situation. The engineering issues can be stated with some certainty, but it is not these issues that will drive the selection. Much will depend on forces outside the control of the engineer who would do well to provide a solution of the form "If *x* then *y*, but if not *x* then *z*." The engineer should also seek the support of the finance and strategy departments to develop a detailed engineering case and cost-benefit analysis and to understand any strategic implications. Armed with these tools, the decision should become more straightforward.

14.3 DECT versus PHS

14.3.1 Background to the debate

The two cordless technologies of DECT and PHS were introduced in Chapter 5. Most major manufacturers have decided to make one of these two products, but not both. Typically, the European manufacturers have selected DECT, while the far-Eastern manufacturers have selected PHS. Both sets of manufacturers have formed groupings typically called MoU groups that they have used to attack the other party. As always, the debate has become vociferous.

14.3.2 The key issues

Different manufacturers have made different claims as shown in Table 14.1.

Table 14.1
Claims Concerning DECT and PHS

Characteristic	Ericsson [2]	Fujitsu [3]
Capacity	DECT 30% greater than PHS	PHS is greater than DECT
Cost	DECT less than PHS	PHS is less than DECT
Planning	DECT easier than PHS	—
Data rate	DECT can provide ISDN whereas PHS cannot	—
Repeaters	Can be provided by DECT but not PHS	—
Interworking with GSM	DECT can achieve this but not PHS	—
Delay	—	PHS is lower than DECT
PSTN connection	—	PHS is simpler than DECT
Power consumption	—	PHS is lower than DECT

Compared to the previous conflict, this would appear to be one that can be resolved more readily by engineering analysis alone. Below, each of these points are addressed in the order in which they are raised.

Capacity DECT provides 12 traffic channels per 2-MHz bandwidth, that is, 6 per megahertz; while PHS provides 4 traffic channels per 300-kHz bandwidth, that is, 13 per megahertz. However, PHS requires around 3 dB more cochannel protection as a result of its more complex modulation scheme, increasing the cluster size that must be used. The effect of this on the cluster size can be modeled as follows. Making use again of Lee's equation, the cluster size of DECT, with an SIR of around 9 dB will be 2.3, whereas that for PHS with a 3 dB greater SIR requirement will be 3.2. Factoring this into the capacity results in a DECT capacity per cell of 2.6 voice channels/cell/MHz, while PHS achieves 4 voice channels/cell/MHz. Hence, it would appear that PHS can provide a greater capacity per megahertz per cell. However, it is not clear whether this is an important issue.

Infrastructure cost It is not generally possible to evaluate the relative infrastructure costs until quotations from the manufacturers have been received. Because fewer DECT transceivers are required since each DECT transceiver can provide 12 channels compared to the 4 of PHS, it may be that economies can be realized in the manufacturing, which will cause DECT to be cheaper. However, in the case that only 4 channels are required, the DECT system will probably be more expensive. In practice, the base station costs will be dominated by the economies of scale achieved and not the manufacturing difficulties. Moreover, in this area, it remains to be seen which technology will achieve the greatest economies. This would appear to be an issue that cannot yet be resolved and depends on the deployment, but where high capacity is required, the greater number of channels on a DECT transceiver would make it likely that DECT would be less expensive.

Simplicity of planning Since both systems use DCA, in most deployments it would seem likely that there is little difference between the two systems. The only slight difference is that PHS requires separate control channels whereas DECT uses the same frequencies for control and traffic channels; as a result, DECT may be slightly simpler to plan.

Provision of 64-kbps bearers It is clearly correct that DECT, with its capability to concatenate channels, can provide higher data rates than PHS. Concatenation of channels will probably be introduced to PHS at some point, but the lower number of bearers per channels means that PHS will always be inferior to DECT in this area.

Provision of repeaters Although there may be repeaters in the DECT product range and not in the PHS range, there seems little reason why PHS repeaters could not be provided if required.

Interworking with GSM The only reason why DECT has an advantage in this area is because standardization has been taking place for some time. If required, PHS could also be made to interwork with GSM, although there may be political difficulties in writing the standards since ETSI is unlikely to welcome the non-European standard.

Delay, connection, and power consumption The specifications show that PHS has a shorter delay than DECT. However, in both cases the delay is insignificant. Both are probably simple to connect to the PSTN and the level of power consumption is unlikely to differ much between them.

14.3.3 Analyzing the debate

In this case, most of the issues can be resolved by engineering analysis, although probably neither of the two parties would agree fully with the preceding analysis, and further study may be more appropriate in some areas. Nevertheless, it seems clear that PHS provides a higher capacity while DECT provides a wider range of services. The major unknown is the relative costs between the two systems. As previously mentioned, since most network decisions are taken on a cost basis, the decisions are difficult to resolve in general. However, in specific cases, with quotes from manufacturers and a network plan, it becomes entirely tractable to resolve the issue.

The complete wireless professional would determine whether capacity or functionality was of overriding importance in this deployment, determine the cost differences between the two systems, and make an informed decision as to the most appropriate choice. Typically, with cordless systems, capacity, although important, is not critical because of the inherently high-capacity nature of the deployment (although this is not the case for WLL systems where cordless cells are larger). Functionality may be critical for certain deployments where marketing requires a particular facility and this should be established. Again, the complete wireless professional will make a decision mostly based on financial rather than technical issues.

14.4 CDMA versus TDMA

14.4.1 Background to the debate

The two cellular systems, GSM and cdmaOne, were introduced in Chapter 3. What was not mentioned was the bitter debate that has taken place between the manufacturers of the two different technologies. The history started around 1991, when a relatively small U.S. company, Qualcomm, announced that it had designed a new mobile radio system based on CDMA technology that offered substantial capacity gains over existing systems. Qualcomm made dramatic claims for their technology, stating that it could provide a capacity over 20 times greater than the existing U.S. AMPS analog technology. Obviously, if true, this was good news for operators and for Qualcomm, but bad news for manufacturers who had invested heavily in GSM, especially as Qualcomm made it clear that they held all the key patents required for their CDMA system.

A parallel debate took place in the United States between cdmaOne and IS-136. In this case the protagonists were Ericsson for IS-136 and Qualcomm for cdmaOne. The debate was very similar to the GSM–cdmaOne debate and is not discussed further in this section.

At first engineers attempted a mathematical analysis and an analysis using simulation to determine whether these claims were correct. However, although both approaches could be used, they required major simplifications that were soon challenged as being unreasonable. Qualcomm took a high-profile approach to tell everyone how good their technology was compared to other technologies which led the GSM manufacturers to attack their claims. The resulting "battle" was a little pointless because it was clear that the answer would not be known until CDMA systems were deployed on a large scale. Nevertheless, it was important to convince a number of operators to do this, and so the "marketing" continued from both sides.

14.4.2 The capacity of CDMA versus TDMA

TDMA and CDMA are both ways of dividing up the radio spectrum so that a number of users can talk at the same time. There are two key issues. The first is the efficiency of the division process taking into account guard bands that need to be provided between adjacent users to stop them interfering. The second factor is the capability of the access method to distribute the interference evenly across all mobiles, as mentioned in Section 2.4.8. Only if the interference is evenly distributed across all the mobiles can the minimum cluster size be adopted and the maximum capacity achieved.

When comparing CDMA and TDMA systems, it is important to assess:

▶ How much spectrum is lost in the process of division of the frequencies?

▶ How successful is the access method in distributing interference evenly across all the users?

Division of the frequencies   Inefficiencies in TDMA result for several reasons.

▶ As discussed in Section 2.4.8, there is a need to allow mobiles time to increase their power at the start of the burst.

▶ Because of the high bandwidth used, ISI becomes problematic and space needs to be set aside for a sounding sequence for the equalizer.

▶ Framing information is required by the mobiles so that they are able to transmit in the correct place.

Of these, the first two are most important. As seen in Section 3.2, in a typical burst of 148 data bits, only 116 can be used to transmit user's data, representing an efficiency of only 78%.

In CDMA, the inefficiency is caused by signals not being received with equal power and hence interfering to a larger extent with other users, reducing the capacity. As shown in Section 2.4.8, with a relatively small error in power control, the system capacity might fall to only 40% of what it would be with perfect power control, although this is something that needs to be proved in actual deployments.

Distributing the interference   Because in CDMA all mobiles use the same frequencies, they interfere with each other to the same extent. As a result, CDMA perfectly distributes interference between users.

TDMA inherently results in uneven interference since some users on the same timeslot will be closer than other users and, hence, will experience unequal interference. As a result, the CDMA cluster size is around one-tenth that of a typical TDMA system. The GSM cluster size can be reduced using some of the techniques described in Section 9.4, which can reduce the gains of CDMA, but this simple analysis suggests that, taking the detrimental effects of power control and the positive effects of interference distribution, CDMA might have a capacity gain of around 2 to 4 times that of GSM.

The complete wireless professional, realizing that a mathematical or simulation approach is intractable, would highlight the key issues and realize that CDMA will typically have a higher capacity than TDMA, with the extent of the advantage depending on the way that both of the networks are deployed and the capacity enhancement techniques adopted within GSM. He or she would wait until networks had been deployed, if possible, to look at the practical evidence, which appears to show that

CDMA typically provides a capacity gain of approximately two to three times that of GSM.

14.4.3 Other issues introduced into the debate

Although the debate tended to focus around which system had the greatest capacity, there were a number of other factors that were also claimed to be important. These included claims that:

▶ CDMA systems have a greater range than GSM.

▶ CDMA does not need frequency planning and, hence, is simpler to deploy than GSM.

▶ CDMA cannot be deployed well in microcells.

▶ CDMA systems have a higher risk.

▶ CDMA systems have a higher cost.

▶ Americans should adopt CDMA as Europe had mandated GSM but were trying to sell GSM into the United States, representing unfair competitive practice.

▶ GSM is deployed worldwide, resulting in roaming capabilities, and the worldwide acceptance makes it the dominant standard.

▶ CDMA systems are difficult to design due to the "cell-breathing" problem.

Greater range   The claim for greater range for CDMA was made because the gain achieved by the despreading process allows the receiver to operate at a lower signal level and hence at a greater distance from the transmitter. While this is true, in heavily loaded systems, the interference from other mobiles and from neighboring cells is such that the range is reduced and CDMA only achieves a similar range to TDMA.

Lack of frequency planning   Because the same frequency is used in each cell within CDMA, there is no need to make a decision as to which frequency to allocate to which cell, unlike TDMA. However, there are now many automated planning tools that can rapidly perform the TDMA frequency allocation process; hence, this is no longer a serious handicap.

Macrocell/microcell   If a microcell is deployed under a macrocell in a CDMA system, then it must use a different frequency. Otherwise, the transmissions made by the mobiles in the macrocell will be at a much greater power level and will cause excessive interference to the microcell. However, if the number of channels required in the microcell is small, then the assignment of an entire frequency may be wasteful. As a result, using a hierarchical cell structure may reduce the efficiency of CDMA.

Risk   Until recently there was a relatively high risk associated with deploying a CDMA system because no other system had been deployed and the problems that might arise were not well known. However, there are now more than 100 CDMA networks around the world and the risk has rapidly fallen to a level similar to that of a TDMA system.

Cost   As with the other conflicts described in this chapter, cost is difficult to compare until there are quotations from manufacturers for identical networks. What appears to be the case is that CDMA base stations are more expensive than TDMA base stations but have a higher capacity and so fewer are required. This generally causes CDMA to be less expensive in a high-capacity situation, but relative cost must be determined for a specific case.

Open markets   There is much discussion in mobile radio as to what extent markets are open around the world, allowing equal competition for all standards. This is a complex political issue rather than one that can be resolved by analysis, but nevertheless, one of which the engineer should be aware.

Roaming   For some users, the roaming that GSM offers will be an advantage. However, in a typical network only around 5% of the users roam, so this is not a key issue. The issue of the dominant standard is only important if the lower economies of scale push up CDMA prices too high, or worse, result in the system eventually being withdrawn from the market. However, given the number of manufacturers involved in CDMA products, the latter seems unlikely.

Breathing problems   Deployment of CDMA systems has shown that there can be problems with cells "breathing." As a CDMA cell becomes more heavily loaded, the interference increases. This results in a reduction in the link budget (because the effective noise floor is now higher) and a corresponding reduction in range. As a result, the size of cells can fluctuate with load. This tends to make cell planning problematic and often results in the system having to be designed so that there is substantial overlap between cells. This ensures that even when cells are contracted, there is still good coverage available.

14.4.4 Analyzing the debate

There appear to be several key issues.

▶ Resolving the capacity difference between the two systems is not possible until deployments have taken place. However, it would seem likely that the capacity of CDMA systems is greater than that of GSM.

▶ It is only possible to determine the relative costs of the two systems for a particular deployment at a particular point in time, but it seems likely that generally CDMA will be more expensive than GSM due to the increased complexity of the system.

▶ There are emotive arguments that cloud the issue.

▶ The large amounts of money and reputations at stake mean that any rational discussion is unlikely.

The complete wireless professional would have realized early on that general arguments about capacity and costs were pointless but that in situations where capacity was essential, CDMA was more likely to be appropriate, whereas in situations where cost was essential, GSM was more likely to be appropriate. The key step is to determine for a particular network design what the different costs of the two systems would be and to use this as a starting point for the analysis. The complete wireless professional would have avoided arguments concerning capacity or general arguments concerning costs as far as was possible.

14.5 Handling conflict

The examples presented here have, perhaps, provided a rather idealized view of the manner in which conflict can be treated. For an engineer in a manufacturer who is committed to a particular product, it may be difficult to view the debate dispassionately; in any case, the employer expects the engineer to support their product. As a consultant pressed for a view as to which of two competing technologies is best, the answer "it depends" does not seem to provide the certainty for which the client is looking. However, there are a number of important lessons that can be learned from this.

▶ Where there are engineering issues, some can be solved using analysis and skill, but there will remain insoluble issues that should be recognized as such. A deep understanding of the underlying mechanisms might help provide some guidance as to the likely solutions to these problems.

▶ In most cases, the key to the problem will be a financial appraisal that may also include a cost-benefit analysis, but to be of any value this will typically need to be for a specific situation.

▶ The complete wireless professional will also need to cope with financial issues, strategic issues, marketing and policy issues, and a number of unknowns. These should be clearly identified and discussed with relevant experts.

The complete wireless professional will do a much better job of resolving conflict than one with a more narrow focus. With the ability to understand most or all of the relevant issues, he or she can put the engineering problems into perspective. Such an engineer will be very valuable in helping clients or employers find their way through the increasing uncertainties in the world of mobile radio.

References

[1]   Lee, W. C. Y., "Spectrum Efficiency in Cellular," *IEEE Trans. on Vehicular Technology*, Vol. 38, No. 2, May 1989, pp. 69–75.

[2]   Akerberg, D., "Comparison of DECT and PHS," Ericsson Report TY95:077, Jan. 1996.

[3]   Hamano, T., "PHS: The Technology and Its Prospects Outside Japan," Mobile Communications International, Nov. 1995, pp. 54–56.

Management

Alice asked, "Cheshire-Puss, would you tell me, please, which way I ought to go from here?" "That depends a good deal on where you want to get to," said the cat.

Lewis Carroll

15.1 Introduction

The complete wireless professional needs to have good managerial skills. He or she will need to be able to manage and motivate staff and be knowledgeable about human resource management and many other similar disciplines. He or she will play a role in developing corporate strategy, particularly in the areas where technology is a key input. This section concentrates on the development of strategy since it is arguably one of the most important skills and the one that lends itself best to textbook analysis as opposed to human resource skills that are better taught in workshop settings. As discussed in more detail in Chapter 16, the complete wireless professional would often benefit from earning MBA in order to have a better understanding of the workings of management. A good overview of all management topics can be found in [1].

15.2 An overview of management

A complete manager would need to have an understanding of the following areas.

- *Management theory* covers areas such as the design of the organization and some of the classic management theories relating to the division of labor and the extent to which managers should be concerned about their staff. The complete wireless professional will typically not need to become overly involved in this area.

- *International management* covers issues associated with international trade, including an understanding of the different cultures; why companies become global; and the different politics, economic, and technical factors in other countries. Although not essential, the complete wireless professional might find it helpful to have some appreciation of the international dimension.

- Unless setting up their own company, *entrepreneurship* is not an area of which the complete wireless professional needs to be particularly aware; however, many organizations now value individuals who can show entrepreneurialism while within the company and this may be something the complete professional may wish to develop.

- *Planning and strategic management* covers the way in which decisions affecting future areas and products for the company are made and is important for the complete wireless professional to understand.

- *Strategy implementation* covers the practical means of putting a new strategy into action, making sure it is accepted by all the staff and that everyone is working toward the new goals. When in a management role, this will be one of the tasks of the complete wireless professional.

▶ *Decision making* looks at the more scientific ways to make a decision between one project and another or between different ways of investing money. Most engineers are intuitively good at this anyway because it typically involves logic and the application of probabilities to outcomes.

▶ *Organizational structure and behavior* covers the division of the work, the departmentalization and span of management, functional and matrix organizations, and ways to go about designing an organization. It is not something in which the typical complete wireless professional will be heavily involved, but a passing understanding of this area may be useful.

▶ *Authority, delegation, and decentralization* covers the nature of authority, the delegation of power to line staff, the design of jobs, and means of enpowerment. As before, an understanding of this area may be useful to the complete wireless professional.

▶ *Human resource management* covers areas such as recruitment, selection, orientation and training, appraisal, promotion, and disciplinary action. It is something in which the complete wireless professional may get involved if they build a team around him or her but hopefully something in which he or she will have support.

▶ The management of *change* is a complex issue involving a need to get employees to accept new ideas and change the manner in which they are working. This is not typically something with which the complete wireless professional would need to be involved but about which he or she should understand the principles.

▶ *Leadership* covers the attributes of the people with power in the organization and looks at a range of different leadership styles. Exceptional leadership skills will not be an essential characteristic of the complete wireless professional, but some leadership skills will be helpful in his or her role.

▶ *Finance and accounting* looks in more detail at the areas introduced in Chapter 8.

▶ *Information systems* covers the design and use of management information systems and will cause few problems for engineers accustomed to computer technology and statistical analysis.

The remaining part of this chapter focuses on the important topic of strategy.

15.3 Understanding corporate strategy

The need for strategy is epitomized by the quote at the top of this chapter—without a firm plan as to where an organization is going, it is likely not to get anywhere. Many organizations seem to try to go somewhere without understanding where, and the result is general confusion and rarely success in the marketplace. The development of strategy is a relatively new concept—most organizations in the sixties and seventies had little idea of where they were going unless they happened to have entrepreneurial leadership. It was not until the 1980s that the science and practice of corporate strategy was developed, mostly by one man, Michael Porter, widely recognized as the giant in this area. All textbooks on corporate strategy, such as [2, 3], reference Porter widely. Actually, Porter's own books are quite hard work, and it is better to read others' interpretation of his words.

Corporate strategy is all about beating your opponents by devising and implementing a strategy that will allow you to sell more products or services than the competition. To some extent this arena could be compared to a game of chess. In the 1960s and 70s, this game was being played with little understanding of good strategy and with ill-defined rules. In the 1980s, some companies started to understand what Porter was saying and became much better players. They won the next games of chess easily and went on to the become the great organizations of the 1980s, those featured in [4]. By the 1990s, most surviving organizations had assimilated much of the knowledge on strategy and the game of chess was again between equal opponents but played on a higher level. Any organization that does not understand strategy will come to the chess board ill-prepared and will be soundly beaten by better opponents. The best way to understand strategy is to read one of the excellent books referred to previously. All the remainder of this section does is to provide a summary of some of the key issues.

Strategy starts with an analysis of the situation facing the company, including all internal and external factors. From this understanding

comes a selection of possible strategies that can then be a
mine which is the most appropriate. Key is the correct
the current situation—from this, potential strategies see
The following are tools to help understand the situatio
cally used in a brainstorming session where a team of people take each
heading in turn and list every factor they can think of, however apparently stupid, in order to try and capture all the issues.

PEST and SWOT   These are acronyms that help the strategist remember to analyze a range of different areas. PEST stands for political, economic, social, and technical.

> ▶ *Political:* An understanding of political changes (such as a change of
> government) or trends (such as deregulation) that are likely to have
> an impact on the sector in coming years; for manufacturers this
> needs to be understood for a number of different countries.

> ▶ *Economic:* An assessment of the economics of the country (such as
> GDP projections) and the economics of the users (such as whether
> operators will be able to borrow money) allowing the ability to pay
> for a product to be assessed; for example, a projection of high GDP
> growth would suggest richer citizens who may be prepared to spend
> more on mobile communications.

> ▶ *Social:* Relevant social factors, such as a backlash against the use of
> mobile phones in public places, or an increase in crime requiring
> mobile phones for security.

> ▶ *Technical:* A look at technical developments that might change the
> market; this includes major breakthroughs such as new batteries,
> general trends such as small and cheaper handsets, and develop-
> ments in related industries such as the capability to send very high
> data rates down fixed lines.

SWOT stands for strengths, weaknesses, opportunities, and threats.

> ▶ *Strengths:* The things that the company does well or has as attributes,
> such as a well-known name, excellent distribution channels, or
> good financing capabilities.

> ▶ *Weaknesses:* The opposite factors such as a lack of skilled staff or an
> outdated manufacturing plant.

▶ *Opportunities:* Areas the company might enter or new products they might develop, such as adding a WLL portfolio based on existing cellular products.

▶ *Threats:* Things that competitors might do that would damage the company such as introducing a much cheaper product or one that is technically superior in a manner that will bring benefits to the users.

Five-forces model   This model was developed by Porter to help people assess the state of their industry. Porter's model is shown in Figure 15.1.

Like the acronyms, the intention is that the strategist considers each of the following different areas.

▶ *Threat of new entrants* looks at whether new companies might enter the market. To understand this it is necessary to consider the barriers to entry; for example, a new cellular equipment manufacturer requires expensive manufacturing plant and new operators require spectrum, which can be difficult to get.

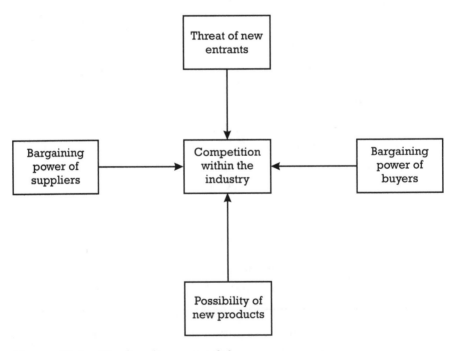

Figure 15.1   The five forces model.

▶ *Bargaining power of suppliers* considers whether the suppliers will alter the way in which the company works by increasing prices, insisting on certain strategies, or favoring a competitor.

▶ *Bargaining power of buyers* is similar for the buyers. Buyers may be able to negotiate reduced prices or changes to equipment. This is especially true in the manufacturing industry where cellular operators are able to negotiate excellent discounts for large orders.

▶ *Competition within the industry* looks at what the existing competitors might do based on knowledge of their current strategy and looks at "what-if" scenarios in the same manner that a chess player will examine all viable moves his or her opponents might make.

▶ *Possibility of new products* tries to look very widely at what might come along in the future. Classic examples are glass bottle makers who failed to see the advent of plastic packaging. To look at what might happen requires a very broad mind, but suggesting a number of possibilities such as wristwatch communicators[1] and then looking at their likelihood and implications is an important exercise.

The value chain   This is another concept developed by Porter, showing that each part of the company contributes to the "value" added to the end product. In making the company as efficient as possible it is important to assess each part of the value chain and ask whether the operation could be outsourced and how the efficiency of the operation could be graded. A possible value chain for a cellular operator is shown in Figure 15.2.

The value chain also shows the interrelationships between different parts of the company. This may allow areas of shared expertise to be identified and, if the company has other divisions, allow sharing of some areas of expertise between the different divisions.

Having used these tools, the situation facing the company should be clear. Strategy development then follows through more brainstorming, development of spreadsheet models, research on particular points raised, and good ideas from those present. Devising a strategy, like inventing a new product, is not something that can be done by following rules but

---

1. Many mobile phone manufacturers are striving to devise the wristwatch communicator and a number of prototypes have already been exhibited. To build such a device requires a shift in thinking away from high-power transmitters, keypads, and speakers integrated into the phone. The development of such devices may change the power base within the world of mobile radio.

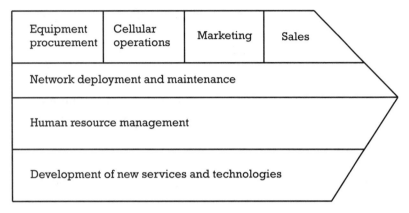

Figure 15.2 A possible value chain for a cellular operator.

requires creativity coupled to a good understanding of the situation. In devising strategy the following points are relevant.

Price or quality   Porter, looking at different companies, said that basically a company can compete on price and become the lowest cost producer; or can compete on quality, producing a product for which customers are prepared to pay more because of some differentiating feature; or can focus on a narrow segment of the market that is so small that other companies will not be inclined to enter the market. He said that companies should follow one of these three strategies—to try and do more than one, or to follow none actively, was to get "stuck in the middle," leading to a lack of direction and eventual failure. Companies that decide to compete on price must have the lowest cost production of all their competitors, otherwise they will enter into a price war that they will lose. Such companies focus on being "lean and mean" with minimal overhead and new techniques for cutting costs—direct line insurance companies who removed the cost of shops in town centers are a good example of this. Companies that compete on quality should have some attribute that it is difficult for others to copy, such as the capability to produce the smallest phones on the market, and need to work continually to maintain this advantage since others will seek to copy. However, while they have a clear differentiator, this can be used to maintain high prices and high profitability—Ferrari is a good example where the name, image, and engines developed in Formula One racing provide a combination of features that others cannot easily copy. Finally, companies that look for a small niche should make sure that they have a product that it is not worthwhile for a large company to copy.

When devising strategy it is important first to decide whether the company will compete on price or quality. This fundamental decision then has implications for the detailed strategy adopted to achieve the target point on the grid. A much more detailed discussion of this technique can be found in [2].

Necessary outputs—the mission statement and the statement of strategy Once the strategy is decided upon, it is important to implement it. Implementation and change management is a large subject, probably outside the scope of the complete wireless professional's expertise. However, the complete wireless professional will want to help in the first stages of communicating the new strategy. These are conventionally the much derided mission statement and strategy documents.

▶ *The mission statement* should be a short, concise statement of exactly what the company is trying to achieve. It should make it clear what are the most important factors. For example, one manufacturer might, as a mission, say, "We will produce the lowest cost cellular handsets on the market," whereas another might say, "We will produce the smallest and most innovative handsets on the market," while another might say, "We will produce the widest range of handsets to fit all the uses and all the markets." The last statement would be a poor strategy because of the attempt to do everything rather than concentrating on price or differentiation. Drafting the mission statement often takes some time, but at the end everyone should be clear on the single most important goal.

▶ *The strategy* should indicate how the goal stated in the mission statement will be achieved. For example, if the mission is to produce the lowest cost handsets, then the strategy might be to work with suppliers to cut costs and to move production to the lowest cost countries. The strategy should be a list of clear steps that will be followed to achieve the required objective. Anyone reading the strategy should be in no doubt that if implemented successfully the strategy would result in the mission being realized.

There is much more than this to strategy. For example, one of the strategies might be to attack a competitor. The attack may be on the competitor's strengths to discredit them, on their weaknesses to quickly gain market share from them, on many fronts simultaneously, on segments where the competitor has little presence, or in "hit and run" attacks where special offers for a limited period take market share before the operator has time to react.

Developing strategy is a big risk business. An inappropriate strategy could ruin the company. The complete wireless professional would be advised to cut his or her teeth on the numerous case studies that are available from management schools. Even better is to complete an MBA where he or she would be guided through the development of strategy in more detail.

References

[1]   Stoner, J., and R. Freeman, *Management*, London: Prentice-Hall, 1992.

[2]   Thompson, A., and A. Strickland, *Strategy Formulation and Implementation*, New York: Irwin, 1992.

[3]   Johnson, G., and K. Scholes, *Exploring Corporate Strategy*, New Jersey: Prentice-Hall, 1993.

[4]   Peters, T., and R. Waterman, *In Search of Excellence*, New York: Harper and Row, 1982.

CHAPTER

16

Contents

The complete wireless professional

Personally I'm always ready to learn, although I do not always like being taught.

Winston Churchill

16.1 Introduction

The first fifteen chapters of this book, in general, provide an overview of the areas of knowledge required to fully understand the world of mobile radio and partake in the conflicts that will undoubtedly occur in the future. This chapter turns to the individuals themselves and to the tools they can use to advance their standing, to enhance their capability to intervene in debates, and to promulgate their opinion more widely than would otherwise be possible. This chapter looks at the topics of conferences, publications, research, and qualifications for the complete wireless professional.

16.2   Conferences and publications

Conferences and publications are useful vehicles for the complete wireless professional to use to publicize his or her analysis of particular situations and to get a wider view of the issues and problems involved in a particular situation. By appearing at conferences, the complete wireless professional will also enhance his or her own standing, which will be useful for a number of reasons:

> Making others more receptive to listen to his or her viewpoint because of the kudos that speaking at conferences can provide;

> Making it easier to approach key decision makers;

> Increasing the wireless professional's visibility, which will help his or her future career prospects.

Getting invited to speak at the conference   Getting invited to speak is relatively simple. Most conference organizers spend a lot of time searching for appropriate speakers. Simply by contacting them with a list of topics on which you would be happy to talk will normally result in a flood of invitations to speak at forthcoming conferences. Getting to know the conference organizer and meeting with him or her occasionally can even result in taking part in some of the preliminary planning of the conference, providing an opportunity to shape the conference along the lines that are personally most relevant. Conference organizers' contact details can be readily obtained from the back of conference advertising material.

Preparing the presentation   Preparation is a task that becomes much simpler with experience. Sufficient time needs to be set aside for preparation—those talks that have been prepared on the train on the way to the conference are generally easy for the audience to identify because they lack structure and thought. However, presentations can also be overprepared. In particular, writing out the complete talk is widely acknowledged to result in poor presentations due to the tendency to then read the presentation rather than speak to the audience, resulting in wooden and insincere presentations. The best form of preparation is to:

1. Make a list of the key points to be delivered during the presentation.

2. Check on the length of time allocated for the presentation and take off 5 minutes for questions and overruns from previous speakers.

3. Produce a first draft of the slides for the presentation, assuming that each slide should take around 2 minutes to present (over time individuals will get to know how long they tend to talk per slide). Make sure that the size of the text is sufficient to be read from the back of a large hall and the background does not reduce its legibility.

4. Go to a closed room and talk aloud through the presentation, timing how long it takes. During this rehearsal make notes on the slides where topics did not seem to flow or where more prompts might be required to remember the key points.

5. Reduce the length of the presentation until it is 10 minutes less than the length of time assigned (so that even allowing the 5 minutes for questions the presentation will still be 5 minutes short). This is a very useful tactic that helps prevent the standard problem of talking too fast during the presentation. Knowing that the presentation is too short tends to result in speakers talking more slowly to avoid finishing too early. A presentation that is too long is generally disastrous for the audience.

6. Redraft the presentation based on the rehearsal so that the flow of the presentation is clear and the audience will easily understand the starting position, the assumptions made, the logical deductions, and the conclusions. Presentations that flow clearly from slide to slide are generally well received.

7. For those who are more experienced at presentations, at this stage consider whether there is any way that the start of your presentation can be made a little different, such as by using a striking graphic or a key statement on the first slide rather than the expected "Contents" slide. This helps grab the attention of the audience. For example, a first slide stating "The market for TETRA

is only one-fourth that of current assumptions" will generally work better than a contents page with the bullets: Introduction, TETRA, The Market, Conclusions.

8. Re-rehearse the presentation, timing it again to ensure that it is still too short. (Even if the presentation really is too short, most audiences prefer this to overly long presentations that impinge on their lunch breaks.) Again check that there is a clear structure and flow to the presentation.

9. If necessary, continue to revise and rehearse the presentation.

With more experience, it becomes possible to shorten this procedure to typically a single rehearsal and often only one draft of the slides. It is important to remain very critical of one's own presentations. Colleagues will typically always say that the presentation was "great" because that is human nature. After every presentation it is worth asking:

▶ Did the presentation flow well while it was being given, or did it have the feeling that it was jumping from topic to topic?

▶ Was it presented to time?

▶ Did the audience appear restless?

In preparing the slides it is important not to overcrowd them with detail (many presenters end up apologizing for "busy slides"). A tendency is to add graphics to every slide on the basis that these will somehow grab the attention of the audience. In practice, simple slides, with graphics only where they really add value, are much more appropriate since they result in the audience focusing on the key points. It is quite routine for a simple presentation, with no graphics, to be rated more highly by an audience than one comprising complex multimedia material.

Giving the presentation   There is much skill in giving a good presentation, but it is something that can be learned and should be carefully studied by the complete wireless professional. The complete wireless professional should aim to go on a two-day presentation skills workshop, lead by an experienced presenter, where presentations are recorded on video and replayed for analysis. Not only will this help ensure that the skills are learned but will also do much to reduce nervousness. Unfortunately, most engineers are introspective individuals not given to flamboyant presentations and often require training to give a more open presentation. Some of the key points to try to remember when giving the presentation are:

▶ At most engineering conferences most of the other presenters are terrible, so it does not take much to appear to be a good presenter.

▶ Most presenters overrun their slot despite thinking they could never talk for that long.

▶ Talking too fast is a standard problem and one that dramatically reduces the message being delivered.

▶ Looking at the audience is essential—it is best to have some system where it is possible to see the slides while still looking toward the audience (e.g., using a PC to run the projection or by interleaving a paper copy of the slides with the acetates and putting the paper copy on the lectern as the acetates are placed on the projector).

▶ Pausing occasionally and altering the tone and volume of voice will help keep the audience awake, especially in the difficult after lunch slot.

▶ Never read a presentation from a script and try to avoid standing behind a lectern if possible since the latter can tend to place a "barrier" between the speaker and the audience.

When in the audience it is worthwhile to try to analyze what the interesting speakers are doing that allows them to keep your attention. Often it will be the combination of humor, a detailed grasp of their subject, and an incisive analysis of the situation. The latter two are things that every wireless professional should strive toward in his or her presentations. Humor is more difficult—forced humor is often worse than no humor at all and the general advice is only to be humorous if it comes naturally to you.

Becoming the chairman   Chairing a conference is often considered a highly prestigious accolade given to those who are recognized as industry gurus. Actually, this is far from the truth. The conference organizers typically do not know who the gurus really are and, in any case, are easily bribed by the conference sponsors to give their key staff the chairman's role. The complete wireless professional should seek to become a chairman because of the prestige associated with it but should realize that getting to be a chairman is something to be worked at in discussion with the organizers rather than something that will naturally follow from years of good work. A very good, if rather obvious, tactic is simply to say to the organizer that you would like to be the chairman. Deliver a CV showing how you have achieved much and keep trying. Do not be put off by a few failures—in many cases, this will be because someone has paid money to be the chairman.

To many at a conference, the chairman might seem to have an easy time, just introducing the speakers and asking some questions. Chairmen can have an easy time, but doing the job well is exhausting. A good chairman should:

- Provide interesting and stimulating opening remarks.

- Read and rewrite the speaker biographies so they are short and to the point.

- Have a number of perceptive and penetrating questions for each speaker.

- Be ruthless in keeping the conference running on time—most speakers overrun even the longest slot and being late for lunch is what the delegates hate the most.

- Run discussion sessions, seeding the discussion with controversial (but well-founded) remarks to get audience participation and making sure that the discussion does not become a dialogue between a speaker and someone in the audience by suggesting this discussion takes place in the coffee break.

- Help the conference organizer by reading messages and relaying logistics.

- Provide a summary of the conference that extracts the key points rather than just summarizes what everyone says and analyses these points to provide some cogent conclusions.

Most of these tasks have to be performed during the day, while listening to presentations, keeping an eye on the time, and thinking up questions. The role of a chairman is worthwhile but not to be undertaken lightly!

Writing papers   Written papers are also a good vehicle for promoting views (and for self-promotion). There is a wide range of different types of written papers.

- Learned journal articles normally based on research work and discussing some new idea or topic;

- Institute magazine articles that are less mathematical and more tutorial but nevertheless are still "learned";

- Papers for monthly magazines that tend to be much more relaxed in style will rarely have any mathematics and will often not discuss any new ideas;

- Conference papers that should support the presentation given.

The learned journal articles are normally the preserve of the research worker who has been investigating some new topic. These journals are typically only read by academia and research establishments, which will not be where either the complete wireless professional, or the people he or she is seeking to influence, will reside. Although it might be helpful for the complete wireless professional to have written some of these learned journal articles in the past in order to boost his or her credentials, this is typically not where he or she should be focusing his or her attention.

The institute magazine articles (e.g., the *IEEE Communications Magazine* or the *IEE Electronics and Communications Journal*) are somewhat more relevant. They will be read more widely by other engineers and allow the chance to analyze existing situations. They also tend to be more respected than articles in monthly magazines. However, they will only be read by other engineers and, in the case of the analysis of a conflict situation, a wider readership including managers and others in the industry might be more appropriate.

Monthly magazine papers carry the lowest level of kudos but typically the highest circulation and are a very good vehicle for the complete wireless professional to express his or her views on a current situation. Before writing the article, look carefully at some back issues of the target magazine to gain an understanding of the style and length of published articles. Reference the editorial contact information inside the front inside cover and contact them to discuss the possible contribution.

Just like presentations, the key to a good paper is clarity and accuracy. A paper that has a clear starting point and a clear direction and works through in an easy-to-follow manner to some logical conclusions will be much better than most of the other papers published and widely read. Papers that just seek to act as an advertising vehicle should be avoided since these are rapidly glanced through by readers.

One of the important attributes for a complete wireless professional is a good writing style. Sadly, it is very rare to find an engineer who can write well—most are more interested in numerate rather than literate disciplines. The best way to learn to write well is to work for someone who can write well and is prepared to spend the time criticizing written work. If this is not possible, courses on writing technical reports will help, as will reading well-written novels. The key to a good writing style is clarity of thought and language and brevity of words. Carefully reread articles before they are sent to the journal asking:

▶ Are the key ideas clearly brought out?

▶ Could anyone follow the logical steps described here?

▶ Is the text "tight" and unambiguous?

When reading early drafts of papers, views are often expressed that the paper needs "polishing." This is generally unhelpful to the author, since if he or she knew how to polish the paper, he or she would probably have done it already. The underlying meaning is that the paper contains all the material that is required but does not read well. This may be because ideas do not flow clearly and the sentence and paragraph construction is such that sentences often need to be read more than once to understand them. When reading your own paper, try to look for areas where the wording and the structure could be improved in order to make it simpler to understand.

16.3 Links with research organizations

Use of research Keeping in touch with research is essential for the complete wireless professional. The research topics of today often become the practical technologies in use tomorrow. The complete wireless professional will need to be aware of research in order to design new products or services, to be able to resolve conflict, and to understand and predict future trends. The simplest way to keep in touch with research topics is to subscribe to the relevant professional journals (e.g., the IEEE in the United States) and to periodically attend conferences. The only problem can be the difficulty in struggling through academic papers—journals that periodically publish reviews and tutorials can be very helpful in providing a solution; for example, see [1–5]. Engineers in large companies may have access to their own research laboratories, which they can use to investigate particular topics. Intranets may also provide a useful summary of work being performed within the company. Beyond this, the engineer can become more proactive with sponsorship or participation in multi-party research programs.

Sponsorship of university research   If the company does not have its own research department or if the research department does not have the capacity to perform all the work required, then it can be "subcontracted" to universities. Universities often provide an inexpensive means of getting work performed. However, they may do so by harnessing individuals with their own agenda, which is often to obtain a doctorate. These individuals will only wish to work on "major" problems that will provide sufficient material for a thesis and may take many years to complete. Hence, universities are often very cost effective for long-term research topics (such as adaptive antennas) but, depending on the university and the individuals involved, may be less suitable as the work becomes closer to commercial reality. Further, the relative lack of control over the research may make universities less well suited for work of the highest importance to the company, although again, this will vary from university to university.

Another way to use universities is simply as a means for generating new ideas. Often university staff and graduates are highly inventive people who can think in unconstrained ways. As a result they will often generate a myriad of new ideas, a few of which may become useful to the company. The role of the complete wireless professional then becomes one of filtering out the good ideas from all the various ideas suggested. This is a surprisingly difficult role. It is human nature to be skeptical of new ideas, especially those furthest removed from current practice; hence, it is easy to dismiss too many of the new ideas. The complete wireless professional should remember this in-built tendency and should think carefully about each of the ideas, discussing them with colleagues where appropriate. Some companies commission universities to give them research briefs and open days to keep abreast of what the university is doing, often favoring those universities where the professor of communications has a good track record of generating successful new ideas.

National and international research programs   There has been an increasing trend toward the use of multiparty research programs, especially in Europe where the *Research into Advanced Communication in Europe* (RACE) and the *Advanced Communications Technologies* (ACTS) programs, sponsored by the European Commission, have generated many thousands of man-years of research. A program like ACTS invites a consortium to put together a research program and bid for ACTS funding. Typically, the EC will fund around 50% of the project with the remainder coming from the participants. The aim is both to maintain Europe's leadership within mobile radio engineering and to foster cooperation between European companies. The history of such programs has been mixed with some generating excellent results, but in other cases the work falls below the highest standards and can disappoint. The difficulty is that the organizations within the consortium often have different goals and different remits. The academics typically want to publish, whereas the industrial partners want exploitable results that they can patent. In some cases, differences such as these, coupled with some partners who did not take the research seriously, has resulted in the premature closure of projects by the EC monitoring committee.

An interesting case of how some of these programs did not meet their goals was the TDMA versus CDMA arena (see Section 14.4). The EC commissioned two projects, one for each access scheme, to determine the best possible performance of the schemes. The idea was that by comparing the results of the two projects, Europe would be able to make a rational decision as to the selection of the air interface for third generation mobile radio (see Chapter 11 for a description of what finally happened). However, both the projects were delayed and, when finally completed, their results incompatible. Not all research projects fail to live up to promise and some have provided very impressive results, but it is clear that careful selection of partners and project management is required to achieve success.

16.4 Qualifications

It is likely that the complete wireless professional will have a degree in engineering, probably electronic engineering or similar from a good university. This is a prerequisite both to provide sufficient background knowledge to be able to understand the fundamentals of mobile radio systems and to demonstrate sufficient intelligence and application to overcome the challenges that will be faced in the future. For most engineers, the degree may be the only form of qualification. However, there is much to be gained in further qualification.

A Ph.D. is a useful, but not essential, tool. It engenders more respect than would otherwise be the case, which can be useful when making a point of view. Assuming that the Ph.D. was conducted on a relevant topic, it can dramatically improve the understanding of how mobile radio systems work. The best Ph.D. would be one that requires the development of a complete simulation of a mobile radio system. Writing such a simulation is an excellent education in all the detailed aspects of mobile radio system design. During the Ph.D., immersion in the university environment can be helpful. Understanding the way in which research is performed is useful when dealing with research institutions. Universities can be usefully harnessed by industry to perform certain research work at a relatively low cost, and an understanding of the individuals in the university and their aims and aspirations helps in this area. The complete wireless professional is likely to have to read learned papers from time to time in order to assess new ideas, and time spent at universities will make this easier. However, the complete wireless professional will not want to spend too much time in a research environment where his or her much broader skill base would be wasted.

An MBA is also useful. Again this engenders respect and prevents the engineer from being pigeonholed as a "techie." Although the complete wireless professional may not have a large team of people to manage, he or she will be involved in financial and strategic issues; an understanding of these will be very helpful in his or her subsequent career.

The complete wireless professional should not scorn the national electrical engineering institutions (e.g., the IEEE in the United States). Institute membership brings with it increased respect, chartered engineering status, and a useful source of information on new developments in mobile communications. Not only should the complete wireless professional join these institutions, he or she should try to increase his or her rank in the institution as often as possible, passing through senior member to eventually fellow status. Partaking in committees and advisory panels further enhances prestige and provides useful contacts to those in the business of generating new ideas.

The complete wireless professional should be well aware of his or her worth to companies. Engineers who are able to understand the wider financial and strategic issues, are articulate and able to write well, and can resolve conflict are very rare. The complete wireless professional will be in very high demand and able to command a salary level much higher than other engineers. The complete wireless professional should set his or her sights high—he or she will be able to achieve much in his or her career.

References

[1]   The *IEE Electronics and Communications Magazine*, published bimonthly by the IEE, London.

[2]   The *IEEE Communications Magazine*, published bi-monthly by the IEEE, New York.

[3]   The *IEEE Spectrum Magazine*, published bimonthly by the IEEE, New York.

[4]   The *IEEE Journal on Selected Areas in Communications*, published bimonthly by the IEEE, New York.

[5]   The *Bell System Technical Journal*, published quarterly by Bell Laboratories, New Jersey.

Appendix A

Erlang B macro

This macro will work on a Excel spreadsheet. When called with the traffic level in Erlangs and the grade of service as a blocking probability (e.g., 1%), it will return the number of circuits required.

```
' Function to return the Circuits required for a given traffic
' figure and Grade Of Service
Function Erlangb(traffic As Single, gos As Single)
    Dim top As Single
    Dim bottom As Single
    Dim circuits As Single
    top = 1
    bottom = 1
    circuits = 1
    While (top / bottom)  gos
        top = top * (traffic / circuits)
        bottom = bottom + top
        circuits = circuits + 1
    Wend
    Erlangb = circuits − 1
End Function
```

Appendix B

Mandating standards

B.1 Introduction

This appendix takes a look at whether spectrum managers or governments should mandate the use of particular standards when assigning licenses to operate a particular service. This practice is widespread within Europe (e.g., GSM/DCS1800 for cellular, TETRA for PMR/PAMR, CT2/DECT for cordless) but is avoided in most other countries. There are arguments both for and against the practice and these are listed in this appendix.

It is important to note that this is a deep and complex issue around which there is currently probably insufficient evidence to fully resolve the problem. This appendix can only scratch the surface of the problem; much more detailed study is required to fully expose all the issues.

An important issue that impinges on this debate is the degree of "openness" of standards. Some standards are truly open, with internal interfaces well defined and IPR issues controlled so that true multisourcing is possible. Others contain proprietary interfaces requiring significant royalty payments that may prevent effective competition. If standards are not mandated and the market is open to competing technologies, it is important to recognize that the widespread deployment of standards that are not fully open could lead to distorted markets.

In Section B.2 each of the key issues is listed and discussed. In Section B.3 some case studies are discussed, while Section B.4 presents some implications.

B.2   The key issues

In this section the arguments used both for and against mandating a standard are examined. The real impact of some of these arguments will become clearer after examining the case studies provided in Section B.3.

Arguments used for mandating a standard include the following.

Economies of scale   It is claimed that mandated standards can result in much greater sales than would have been the case otherwise, driving down end user prices and hence maximizing the benefits.

Standards certainly provide the potential for economies of scale, the question is whether mandating them improves this. GSM has achieved excellent economies of scale, but arguably this is due to the openness of the standard and its excellent design making it attractive around the world. Other mandated standards have not achieved such good economies of scale. Key issues appear to include the ability of manufacturers to spread research and development over many customers, a common spectrum allocation so that different mobiles are not required in each country, and the ability to sell in most countries, that is, national markets are not barred to standards. It would appear that producing well-designed and truly open standards are the key factors in achieving economies of scale and not the mandating of the standard.

Support to European industry   It is claimed that mandated standards allow European industry to develop products that they are assured of selling within the European market and provide substantial export opportunities.

There are many problems with this argument, which can be summarized by saying that mandating standards may be necessary for success but is not sufficient—market conditions for the acceptance of the standard must also exist. First, mandated standards do not necessarily ensure

European sales—DSRR is a good counterexample to the obvious success of GSM. Even where successful, mandated standards do not guarantee success to the local industry—few of the CT-2 sales benefit the United Kingdom, where the standard first appeared in Europe. Finally, management research by Porter [1] and others has shown that a home market biased toward national organizations tends to hinder rather than aid international competitiveness; nor does using mandated standards to encourage the growth of local industry seem to work despite it being tried recently in Korea and other countries. Most management gurus would argue that the likes of Nokia and Ericsson were world leaders due to their innovation and skills and not due to a home market encouraged to buy their products. Also, non-European companies (e.g., Motorola, AT&T, NEC, and Panasonic) sell widely into the European market and only 5 out of the current top 20 mobile operators are European. The case for this argument looks weak.

European roaming   It is argued that mandating standards across Europe allows users to roam across Europe with a single terminal providing substantial end-user benefits.

GSM has clearly been successful in this regard and the benefits of roaming are valued highly by a subset (around 5%) of its users. Around 8% of potential TETRA users claim to value roaming. Roaming provides additional societal benefits such as enhancing the "borderless Europe" concept through the equal provision of information and services to all countries, enhances the mobility of businesses, and helps with the introduction of the Information Society. At a lower level, roaming overcomes the need for type approval in each country.

To some extent, if roaming is valued by users, then they would be prepared to pay a premium for GSM-compliant systems and hence the market process would provide a roaming facility. However, the societal benefits are unlikely to be achieved by the market alone and mandating of common standards may be necessary in this regard. Further study of the importance of mandating standards to achieve societal benefits would be worthwhile.

Interference   It is claimed that a common standard and frequency allocation reduces interference between neighboring countries and is required for European frequency harmonization.

Cross-border interference is a major issue requiring significant effort on the part of national operators. In principle, a spectrum mask is all that is required to eliminate interference problems as is used in the United States and elsewhere. In practice, detailed planning between different operators across borders is essential. Interference may be particularly problematic if very different multiple access mechanisms are in use on either side of the border (e.g., CDMA and TDMA). However, the use of the same standard does not seem to significantly ease this interference problem and, hence, the need for mandated standards here appears weak.

Maintenance of Europe's lead in mobile radio   By continually developing standards whose use is mandated, it is argued that Europe continues to lead the world in mobile radio.

Arguably, it does not. The most innovative technologies are now arising within countries unconstrained by mandated standards. As mentioned earlier, many would argue the success of European countries comes despite having imposed standards, not because of it. Finally, mandated standards do not prevent non-European countries entering the market, so the success of a European standard may not necessarily bring benefits to Europe.

Mandating a standard provides good spectrum efficiency Since the spectrum manager can have an input into the spectrum efficiency of the standard as it is developed, he or she can ensure that it makes good use of the available resource.

This is not necessarily true. Good spectrum efficiency will only occur if the mandated standard is the most technically efficient available. This may not be so in many cases (e.g., it is argued that CDMA is technically more efficient that GSM). Worse, spectrum pricing, as it is being considered by many countries, encourages efficient use of radio spectrum, but mandating standards removes one degree of freedom for users to improve their efficiency. (Note that changing technical efficiency is necessarily a slow process as operators depreciate their existing equipment.)

Mandating a standard increases the openness of the standard Having standards that are truly open is critical—this allows multiple sourcing of components and terminals, resulting in a highly competitive market. Mandating forces increased openness through a more intensive standards-making process and regulations regarding IPRs of mandated standards.

There is little evidence to support this. Some mandated standards are open (e.g., GSM) whereas others are highly proprietary (e.g., Linear Modulation in the United Kingdom). It appears to be the rules and philosophy imposed by the standards-making bodies such as ETSI that are instrumental in achieving openness and not the act of mandating. Whatever, it is clear that a standard that is not fully open in all respects should not be mandated since this provides a monopoly advantage to the owners of the key IPRs.

Achieving the single European market   The EC has been keen to encourage and mandate the harmonization of technical standards to enable the free movement of goods and services across member states. While open standards may promote free trade, idiosyncratic national standards may act as nontariff barriers (e.g., the French choice of SECAM for its color TV standard).

There is much truth in this. However, what is not clear is whether the costs of opening the European market in this way exceed the costs of having a closed European market in some areas. Possible costs might include the potential adverse impact on innovation. This is an area that merits further study.

Increasing consumer and producer certainty and encouraging the early adoption of new standards   Faced with a proliferation of standards, producers and consumers are often unsure as to which technology to adopt in case they become stranded with an obsolete product or locked into the single supplier of a proprietary standard. This may lead to a waiting game in which players delay decisions until the market becomes clearer. In extreme cases this prevents new technology emerging (e.g., the digital cassette). Government intervention in the form of mandating the use of a particular standard can arguably overcome this problem.

There is certainly some truth in this. Further study would be useful to determine whether this is a key issue or whether having only a single standard (which was not mandated) but was manifestly open would be sufficient. Also, increasing certainty will not help in the case that an inappropriate standard has been selected (e.g., DSRR).

Arguments used against mandating a standard include the following.

Enforced use of standard is contrary to a free market approach and can distort the market   This somewhat idealistic argument is undoubtedly true. Enforced use of a standard occurs in few other areas and inhibits innovation. If the "wrong" standard is selected (e.g., DSRR) the users will suffer. If the "right" standard is selected (e.g., GSM) users will gain little over a free market approach. If the standard is, perhaps, more costly than available alternatives and the service competes with another (e.g., PAMR versus cellular), then the market can be distorted with resources wasted and competition restricted. If other market-oriented approaches like spectrum pricing are in use, the consequences may be significant.

This is undoubtedly true. Economists would argue strongly that the restrictions on users should be limited to a minimum and that distorting markets is a bad thing (the Communist regimes being the example of extreme market distortion). Note, however, that the use of proprietary technologies can distort the market to a greater degree than open standards and should be carefully avoided.

How can the government know best?   To some extent this is the same argument as the free market but asks how government bodies are able to select the optimum standard. This is sometimes known as the "blind giant problem."

This argument is probably false. Government does not design the standard, industry does. Government only mandates a standard developed by industry who have a better claim to understand the market. This approach should continue.

Mandating standards prevents innovation   There are some worrying trends that might be attributable to mandating standards. Many of the innovative ideas in mobile radio now appear to emanate from outside Europe—these include CDMA cellular systems and frequency hopping private radio systems. It is possible that these are early signs that mandating particular standards can stifle innovation. These new systems also counter the view held by many that unless a standard is mandated, thus almost guaranteeing a market, manufacturers will not take the risk of investing in the development. Indeed, despite the closed market in Europe, and hence reduced sales potential, manufacturers outside Europe are still willing to take these risks where they see the potential for volume sales.

Mandating a standard is contrary to free-trade principles
In theory only European organizations (or non-European organizations with a significant European presence) can partake in the ETSI standardization process (non-European organizations do not have voting rights) and, hence, influence the development of radio system for sales into Europe. The European market is then blocked, through a mandated standard, to U.S. equipment, for example. However, Europe expects to be able to export its standards to the open U.S. market. Understandably, U.S. companies can be annoyed by this behavior.

Loss of benefits arising from product differentiation and increased consumer choice   Much of this issue depends on the relative values that consumers place on standardization versus the benefits of product differentiation and product choice.

The strength of this argument depends on whether the European market could support more than one standard. If this is so, imposing a standard may impose additional costs on the consumer relative to the free market. Further investigation is required to determine in which situations more than one standard could be supported, but many would suspect that in most cases only a single standard could be supported.

These arguments are now considered in the context of case studies in the following section.

B.3   Case studies

In this section a number of case studies are considered that help to throw light on the arguments presented in the previous section.

GSM   This is the example that proponents of mandating standards quote when they wish to demonstrate its success. With 198 networks in 92 countries and 100 million users at the time of writing, GSM is a stunning success. Note, however, that even greater success has been achieved by, for example, VHS and analog compact cassette without mandating a standard—they have become de facto standards. Many countries are now selecting GSM because of the competitive supplier market. However, the real question is, "Would the standard have been less successful if its use was not mandated"? While it is always difficult to conjecture as to what might have been, it would appear that even if its use had not been mandated in Europe it would have been widely adopted because, at the time, it was the best system available. So we would argue that standardization was undoubtedly a good idea and mandating its use unnecessary but not problematic since it would have been adopted anyway.

Proponents of mandated standards argue that mandating has led to a European cellular industry so strong that it sells product into a fragmented U.S. market that has spent effort infighting rather than joining forces to attack the world market. We would argue that the U.S. "problem" was a lack of a truly open U.S. standard and not a lack of a mandated standard. Note also that the frequent recourse to litigation in the United States can persuade spectrum managers to license a range of standards.

Mobile data   In the United Kingdom particularly, but also throughout Europe, mobile data networks have largely failed to live up to their initial subscriber estimates. Some would argue that this was because there was no mandated standard. (In fact, there was no standard at all, so it would have been difficult to mandate one.) Others would argue that failure was caused by too many competitors, poor marketing, a lack of focus on user applications, and overly optimistic initial estimates. They would also point out that there now is a de facto standard in Mobitex.

GSM is now starting to develop a significant data market, but this does not necessarily prove that a mandated standard was required to penetrate the market; rather that as an added extra on top of voice, data has more appeal than when offered alone. The argument that a mandated standard would have improved this market appears, with this cursory assessment, to be weak.

Digital short-range radio   This is often quoted as the counterexample to the success of mandating standards. The standard was developed and spectrum allocated for its use, but since no equipment was developed no one was able to offer a service. Now the spectrum has been reallocated to other uses and proprietary technology allowed for this service. If there had been no mandated standards, users could have had a service more rapidly, potentially providing greater economic benefit to Europe. The key problem was the development of the "wrong" standard. This is the fundamental point: mandating a standard runs the risk of picking the wrong one. Here, not mandating a standard would have certainly provided some benefits. (Interestingly, DSRR was developed in reaction to concerns that a Japanese system would dominate the European market and not in reaction to a perceived user need.)

Digitally enhanced cordless telephone   Here the mandated standard has resulted in limited and slow take up of the product. Other, arguably more appropriate, technologies such as PHS have made much more rapid inroads outside Europe. If a standard had not been mandated, then possibly manufacturers may have produced a more appropriate technology, easing introduction and increasing international competitiveness. Again the "wrong" standard appears to have been developed with DECT appearing to be attempting to simultaneously address the public and private markets, preventing it from targeting a particular market well. The CT-2 standard appeared to have a similar fate.

B.4 Implications

This appendix has briefly and superficially considered a wide range of issues. This section tries to tie some of these together to arrive at some conclusions.

Since approximately 1980, European spectrum managers have been mandating the use of particular standards for particular spectrum bands. This practice appears to have occurred largely without question, possibly due to the support of European bodies. Spectrum managers have used a number of arguments to make their case, some of which, when analyzed, appear unconvincing, others of which may have substance and merit further investigation. The arguments against mandating, although generally ideological, appear mostly valid.

Above all, spectrum managers have looked at GSM, which has been a great success, and compared it to the U.S. cellular market and reached the conclusion that mandating a standard provides Europe with a significant competitive advantage. As argued in Sections B.2 and B.3, this view is flawed and attempts to repeat the success of GSM, though mandating additional standards are not guaranteed to succeed. GSM succeeded because there was a strong user requirement for a standard mobile radio system across Europe and the GSM standard met the user requirement. It may have been that it did not need to be mandated to succeed, although this would have depended upon the extent to which individual countries would have tried to develop their own systems. The United States failed not because they allowed a range of proprietary systems to use the radio spectrum but because industry failed to agree on a common standard.

Europe has taken a lead in cellular radio because of its strong research base, early availability of radio spectrum for cellular applications, strong standards-making body, and economic and topographic factors. It is now starting to lose that edge as innovative CDMA technologies are introduced in the United States and as the United States dominates the mobile satellite arena. Its lead in cordless technology is being threatened by the Japanese personal handiphone system, and its lead in paging is under threat by the U.S. Flex system manufactured by Motorola, possibly due to the fact that DECT and ERMES are not well aligned with user requirements. Arguably, mandated standards is reducing the ability of Europe to react to user needs and to provide competitive systems.

A factor that has been noted by many as of overriding importance is the openness of the standard. Fundamentally, this does not appear to be affected by whether the standard is mandated but by the standards-making process. For this reason, openness has not been considered in detail in this appendix.

In summary, it would appear that there are arguments for and against mandating, most of which require further investigation to determine whether they are valid.

B.5   Conclusions

This appendix raises some of the issues associated with the mandating of particular standards when licensing an organization to provide a particular service. The practice of mandating seems widespread within European governments and spectrum managers, despite their often heard claims that they embrace free market ideals.

Of the arguments for and against mandating, the following views can be summarized:

▶ Mandating does not appear to materially affect the achievement of economies of scale; it is the openness and appropriateness of standards that achieves this.

▶ Mandating may actually hinder rather than help local industry.

▶ Mandating does help roaming and may provide important societal benefits.

▶ Mandating may hinder Europe maintaining its current lead in mobile radio by preventing innovation.

▶ Mandating may prevent the most spectrally efficient technologies being adopted.

▶ Mandating may help to achieve a single European market.

▶ Mandating may be required to stimulate a market and prevent manufacturers and consumers waiting for the winning standard to emerge.

▶ Mandating is contrary to a free market approach.

▶ Mandating may decrease consumer choice.

A number of case studies were discussed that appeared to indicate that standardization can be helpful, but for the best advantage, standards should not be mandated. It was also noted that mandating may be reducing the level of innovation within European mobile radio industry, may be preventing the most technically efficient use of spectrum (and possibly reducing the economic benefit to the country), and may be contrary to free trade ideals.

References

[1]   Porter, M., *The Competitive Advantage of Nations*, London: MacMillan Press, 1998.

Glossary

AbS   Analysis by synthesis. A type of speech coding where an attempt is made to model the vocal tract of the speaker in order to reduce the information needed to be transmitted.

ADPCM   Adaptive Differential Pulse Code Modulation. A relatively simple form of speech coding using a differential form of PCM.

ADSL   Asymmetric digital subscriber line. A means of achieving high data rates on twisted copper pair telephone wires and capable of providing up to 8 Mbps from the network to the subscriber and a few 100 kbps in the reverse direction.

Allocation   The process in spectrum management of determining the purpose for which a piece of spectrum should be used.

AM   Amplitude modulation. The modulation of the amplitude of a carrier wave in accordance with the information to be transmitted.

AMPS   Advanced Mobile Phone Service. An analog phone standard in widespread use in North and South America.

AMR   Advanced multirate speech coder. A proposal for a new speech coder for GSM which would vary its coding rate in accordance with the information to be transmitted.

ANSI   American National Standards Institute. The U.S. body responsible for setting standards.

ARIB   Association of Radio Industries and Businesses. The Japanese body responsible for setting standards.

ARQ   Automatic repeat request. An error correction technique whereby blocks that are determined to be in error are resent.

ASCI   Advanced speech call items. The additions to the GSM standard that will provide group and broadcast calls.

Assignment   The process in spectrum management of deciding which user should be able to use a particular piece of radio spectrum.

ATIS   Alliance for Telecommunication Industry Solutions. The U.S. body responsible for the standardization of GSM-based solutions for use in the United States.

ATM   Asynchronous transfer mode. A technique for the transmission of data over fixed networks where the transmitter and the receiver are not synchronized.

AuC   Authentication center. The entity within GSM that authenticates the mobiles and provides a key for encryption purposes.

Autocorrelation   The process of multiplying a sequence by a copy of itself, when the copy is delayed by a variable time.

AVL   Automatic vehicle location. The technique of determining the position of a vehicle.

Back-to-back   Transmission from one mobile directly to another without using base stations.

BCCH   Broadcast control channel. The channel within GSM that provides information on the configuration of the cell.

BCH   Bose-Chaudhuri-Hocquenghem. A type of block coding error correction.

BDT   Bureau for the Development of Telecommunications. Part of the ITU.

Beauty contest   A technique for deciding to whom to assign a license by considering the relative merits of a number of competing bids.

Bent pipe   A term used in satellite systems where any signal received by the satellite is immediately rebroadcast back to the Earth station under the satellite.

BER   Bit error rate. The percentage of bits in error when transmitted through a channel.

BOT   Build, operate, and transfer. The act of a third party building a network, starting to operate the network, and then transferring the network to the owner.

BPSK   Binary phase shift keying. A type of modulation where there are only two possible constellation points and the phase of the carrier is changed in order to move between constellation points.

BRA   Basic rate access (ISDN at 144 kbps). A type of ISDN provision where the user is given two 64-kbps bearer channels and a 16-kbps data channel.

BSC   Base station controller. The entity within GSM responsible for controlling the base stations.

BSS   Base station subsystem. A term for the combination of BTSs and BSCs within the GSM network.

BT   Bandwidth-time product. A measure of the amount by which a pulse is distorted prior to modulation where a low BT product results in a low transmitted bandwidth but spreading of the input pulses over a number of transmitted data symbols.

BTS   Base transceiver station. The entity within GSM responsible for the transmission of radio signals.

Busy hour   The hour during the day when there is the most traffic on the network.

CAMEL   Customized applications for mobile network enhanced logic. An addition to the GSM standard allowing support for operator-specific features for roaming mobiles. To do this, intelligent network principles are being introduced into the network.

CBS   Common base station. The use within PMR of a base station shared between a number of users.

CCIR   Consultative Committee for International Radio. The previous name for the ITU-R.

CCIT   Consultative Committee for International Telegraphy. The previous name for the ITU-T.

CCITT   Central Committee for Telephony and Telegraphy. The previous name for the ITU.

CCS   Common channel signaling. The sending of signaling information on the same fixed line as voice information. The signaling is normally carried out of band.

CCSC   Control channel system codeword. The signaling used within the MPT1327 standard to identify mobiles or groups of mobiles.

CDMA   Code division multiple access. The division of radio spectrum by the provision of a different code to each of the users.

CELP   Code excited linear prediction. A type of speech coding where excitation sequences are selected from a codebook and those most accurately describing the block of speech sent to the receiver.

CEPT   European Conference of Postal and Telecommunications Administrations. The European spectrum manager.

CIR   Committed information rate. The rate that packet-based fixed network protocols guarantee to provide to a particular user.

Clearance   The act of a spectrum manager of removing existing users from the radio spectrum so that it can be allocated to a new use.

Clipping   The loss of the first syllable of each burst of speech due to the delay in the speech activity detector identifying that a user has started speaking.

CMIP   Common management information protocol. The protocol used by TMN.

CT-2   Cordless telephony system 2. A digital cordless telephone system, now generally being replaced by DECT.

CTCSS   Continuous tone-controlled signaling system. A system used in simple PMR radios to identify the group to which the call is destined through the continuous transmission of low-frequency audio tones.

D-AMPS   Digital AMPS. A digital cellular standard mainly in use in North and South America, also known as IS-136 and TDMA.

DCA   Dynamic channel allocation. A system whereby base stations do not have dedicated channels but select a channel when it is needed for communications on the basis of the channels currently suffering the lowest interference. As a result, system capacity is typically increased.

DCS   Digitally controlled squelch. A system used in PMR where the mobile's speaker is turned off unless there is a signal from the network activating the mobile.

DCS1800   Digital cellular system at 1,800 MHz. A digital system based on GSM and operating at 1,800 MHz. This has been renamed GSM1800.

DECT   Digital enhanced cordless telephone. A cordless standard now widely deployed around the world.

DPSK   Differential phase shift keying. A modulation technique where the transmitted information is encoded through the phase difference between the previous and current symbols.

D-QPSK   Differential quadrature phase shift keying. A modulation technique where the transmitted information is encoded through the phase difference between the previous and current symbols and there are four possible symbol points.

DS   Direct sequence. A form of CDMA where the input data is multiplied by a code.

DSL   Digital subscriber line. A technique to enhance the data rates that can be sent down copper lines using advanced modulation techniques.

DSRR   Digital short-range radio. A standard for short-range radio systems providing back-to-back communications, which has been unsuccessful.

DTI   Department of Trade and Industry. Part of the U.K. government.

DUP   Data user part. A protocol within SS7 responsible for the transmission of data.

EBIT   Earnings before interest and taxation. An accounting measure that details the profit before interest and taxation are deducted.

EBRD   European Bank for Reconstruction and Development

EC   European Commission

ECM   Explicit congestion message. A message sent in packet-switched fixed network protocols to warn the information source to reduce the rate of the transmitted data.

EFR   Enhanced full-rate speech coder. A new GSM speech coder working at the same data rate as previous coders but providing a better voice quality.

EIR   Equipment identity register (when used in conjunction with mobile networks). A register within GSM that keeps a record of stolen mobiles so that they can be barred from use on the network.

EIR   Excess information rate (when used in conjunction with fixed networks). An agreed data rate set within ATM networks at which information can be transmitted in the case that the network is not congested.

EIRENE   European Integrated Radio Enhanced Network. The European railway project to provide a digital mobile radio system.

EMLPP   Enhanced multilayer priority and preemption. The addition of priority and preemption capabilities to the GSM standard.

ERMES   European Radio Messaging Service. A standard for digital paging systems.

ETSI   European Telecommunications Standards Institute. The European body responsible for standardizing telecommunication systems.

FACCH   Fast associated control channel. The channel used within GSM to send signaling information to the user in the case that more signaling information than can be accommodated on the SACCH is required. Transmission of FACCH information results in stealing speech frames.

Fade   The loss, or significant reduction, in radio signal in a particular location.

Fast fading   The term given to a mobile radio propagation phenomena whereby the mobile experiences many fades per second. Fast fading is caused by multipath propagation.

FCC   Federal Communications Commission. The U.S. body responsible for the management of radio spectrum for non-governmental use.

FDM   Frequency division multiplexing. The division of the input data stream into a number of substreams that are then sent on separate mobile radio channels.

FDMA   Frequency division multiple access. The division of the mobile spectrum into a number of frequency blocks that users access when they require a channel.

FE   Functional entity. A part within the IN that is responsible for performing a particular function.

FEA   Functional entity action. The action performed by a FE.

FFSK   Fast frequency shift keying. A type of modulation where the frequency of the carrier is changed in accordance with the information transmitted by the minimum amount required to ensure the signal is correctly decoded.

FH   Frequency hopping. The changing of frequency during the process of transmission, often hundreds of times per second, in order to evenly distribute interference and overcome fading.

Free space attenuation   The loss that a signal experiences as a result of traveling through free space. It is proportional to the distance squared.

Fresnel zone   An area around a LOS path that must remain clear if the signal is not to suffer significant attenuation.

GAP   Generic Access Protocol. Part of the DECT specification that ensures mobiles will be able to roam from one network to another.

GDP   Gross Domestic Product. The value of the business conducted within a country.

GMSC   Gateway mobile switching center. An entity within GSM that forms the interface between a GSM network and other networks.

GMSK   Gaussian minimum shift keying. A form of phase modulation where the input data is shaped by a filter having a Gaussian impulse response prior to transmission.

GPRS   General packet radio service. An addition to the GSM standard that allows the transmission of packet data.

GPS   Global Positioning System. A system allowing receivers to locate themselves to an accuracy of 100m or better. The system works by measuring the received signals from a number of satellites.

GSM   Global system for mobile communications. The most successful digital cellular system in the world.

HDSL   High-speed digital subscriber line. A subset of the DSL family where symmetric data rates of up to 2 Mbps are provided using up to three parallel copper pairs.

HDTV   High-definition TV. A form of TV where a higher definition picture than normal is transmitted.

HFC   Hybrid fiber coax. A design technique used in cable networks where fiber optic cables are used in the backbone and coaxial cable in the last mile drop to the subscribers.

HLR   Home location register. The entity within GSM that stores the last known location of the subscribers to the accuracy of an MSC area.

HSCSD   High-speed circuit-switched data. An enhancement to the GSM standard whereby users can concatenate a number of timeslots to obtain data rates of up to 64 kbps.

IEE   Institute of Electrical Engineers. The U.K. professional body representing electronic and electrical engineers.

IEEE   Institute of Electrical and Electronic Engineers. The U.S. professional body representing electronic and electrical engineers.

IMEI   International mobile equipment identity. The number used within GSM to identify the mobile phone that is being used to make the call.

IMSI   International mobile subscriber identity. The number used internally within GSM to identify the subscriber who is making the call. This number is derived from the dialed number, the MSISDN during the call initiation process.

IMTS   Improved Mobile Telephone Service. An early analog phone service deployed in the United States.

IN   Intelligent network. The concept of adding a capability to program a fixed phone network so that new services can be readily added without needing to upgrade the hardware. The concept also includes an architecture design that can realize this capability.

INAP   Intelligent network application part. A protocol used within the SS7 signaling system for the support of IN.

IOU   Intelligent overlay underlay. A technique used within GSM to enhance the capacity of a network by deploying carriers in a cell with different transmit powers and different reuse patterns.

IP   Intelligent peripheral. An entity within IN responsible for gathering information from and supplying information to the user.

IPO   Initial public offering. The first sale of shares in a company onto the stock market.

IPR   Intellectual property right. A patent or similar protective distinguishment that gives the inventor rights to exploit their invention without others copying it.

IRR   Internal rate of return. An accounting measure of the effective interest rate that would have to be available from a bank in order to provide the same return on investment as the project under consideration.

ISDN Integrated services digital network. A network protocol for fixed networks that provides digital transmission for users in multiples of 64 kbps.

ISI Intersymbol interference. A propagation phenomena whereby a signal containing the previous bit, or symbol, transmitted, is sufficiently delayed that it interferes with the current bit or symbol.

ISM Industrial scientific and medical. A category of use of radio spectrum whereby the spectrum is not used for communications but for purposes such as microwave heating.

ISUP Integrated services user part. Part of the SS7 signaling system protocol responsible for carrying ISDN data.

ITU International Telecommunications Union. The worldwide body responsible for spectrum management.

LAN Local area network. A communications network, often fixed, that provides communications between a small group of users in the same building or on the same campus.

LEO Low Earth orbit. A satellite orbit between 500 km and 1,000 km above the surface of the Earth, used by the new generation of personal communication satellites.

Limited mobility The ability of the user to make a call on a mobile phone but only within a restricted area, often covering a single city.

LMDS Local multipoint distribution system. A TV transmission system using radio propagation at around 28 GHz. It is also being proposed for wideband data transmission.

Long code A code used within the cdmaOne system that is assigned to a particular user to distinguish them from other users.

LOS Line of sight. A form of radio propagation where it is possible for the receiver to visually see the transmitter.

LTP Long-term predictor. Part of a speech coder that provides information about relatively long-term variations in the speech waveform.

MAP Mobile application part. Part of the SS7 signaling protocol used to support GSM signaling between BSC, MSC, and location registers.

MBA   Master of Business Administration. An academic degree in business management.

MCCH   Main control channel. The name of the control channel used within TETRA.

MLPP   Multilevel priority and preemption. The name of the priority and preemption scheme used within ISDN.

MMDS   Microwave multipoint distribution system. A type of TV distribution system using analog transmission and working in the 2.5-GHz band, providing one-way transmission only.

MOS   Mean opinion score. The score given to a speech coder by a panel of listeners to rate its ability to accurately encode and decode voice.

MPE   Multipulse excited. A form of speech coder where the excitation waveform is represented by multiple pulses.

MSC   Mobile switching center. The entity within a GSM network responsible for switching all calls.

MSISDN   Mobile station integrated services digital number. A mobile number used within GSM as the number dialed by users who wish to contact a mobile.

MSK   Minimum shift keying. A type of phase modulation whereby the phase of the carrier is only changed by half a cycle, in a linear manner, over a bit period.

MTP   Message transfer part. A part of the SS7 protocol responsible for the accurate transmission of messages across the network.

MTS   Mobile Telephone Service. An early form of mobile radio system deployed in the United States in 1946.

MVDS   Microwave video distribution system. A type of TV distribution system standardized across Europe, working at 40 GHz and providing a large number of TV channels on the downlink with some limited return path capability.

NMT   Nordic Mobile Telephone. An analog mobile phone standard used widely in Scandinavia at 450 MHz and 900 MHz.

NPV   Net present value. An accounting measure used to determine the profit from a project after discounting future cash flows by the opportunity cost of capital.

ODMA   Opportunity driven multiple access. A scheme whereby mobiles who hear a message destined for another mobile can retransmit this message if the other mobile is out of reach of a base station.

OFDM   Orthogonal frequency division multiplexing. A category of FDM where the subchannels used for transmission are orthogonal to one another.

OMAP   Operations and maintenance application part. Part of the SS7 signaling protocol responsible for passing messages to the operations and maintenance center.

OMC   Operations and maintenance center. The part of a mobile or fixed network that monitors the operation of the network for faults and can modify the parameters of remote parts of the network, such as the frequency at which a particular base station transmits.

OnePhone   A concept developed by BT whereby users have a dual-mode DECT/GSM phone that selects the DECT standard when indoors and the GSM standard when out of doors.

O-QPSK   Offset quadrature phase shift keying. A type of QPSK where the change between symbols on one of the quadrature branches is offset by half a symbol period compared to the other branch to avoid the transmitted wave form passing through the origin during a change of symbol.

O&M   Operations and maintenance. Another term for the OMC.

PABX   Public access branch exchange. A switch in a fixed network.

PAMR   Public access mobile radio. A type of PMR where the network is provided by a third party, or operator, and shared between a number of users.

PBX   Private branch exchange. A small switch within an office providing service to the users in the office, allowing interoffice calls, and providing a point of interconnection with the PSTN.

PCM   Pulse code modulation. A simple voice-coding technique using regular samples to provide a digital datastream at 64 kbps.

PCS   Personal communications service. Another name for digital cellular operators, particularly applied to operators in the United States with spectrum at 1,900 MHz.

PDC   Personal digital cellular. The Japanese digital cellular standard.

PDH   Plesiochronous digital hierarchy. A technique used in fixed networks to communicate between switches running at different clock speeds by a process of multiplexing and demultiplexing.

PEST   Political, economic, social, and technological. An acronym used by management when conducting brainstorming sessions to set strategy.

PHS   Personal handiphone system. A Japanese cordless standard.

Pilot Tone   A signal sent at a constant power level which allows mobiles to detect the presence of carriers.

PLMN   Public land mobile network. The normal term for a mobile network such as that operated by a mobile radio operator.

PM   Phase modulation. The modulation of the phase of the carrier wave in accordance with the information to be transmitted.

PMR   Private mobile radio. A use of radio systems where the user provides their own infrastructure. There is a complete industry with different mobiles and infrastructure to support this particular mode of operation.

PN   Pseudorandom noise. A type of mathematical sequence with particular properties such that the signal is very close in appearance to random noise.

PRA   Primary rate access. The provision of 30 data channels each providing 64 kbps within ISDN.

PRMA   Packet reservation multiple access. A type of access whereby users only transmit when they have information to send and the network reserves them resource for the duration of their transmission.

PSTN   Public switched telephone network. The normal name for a fixed telephone network such as that operated by the PTO.

PTO   Post and Telecommunication Organization. The normal term for the company that operates the main PSTN within country. The PTO is state-owned in some countries.

QAM   Quadrature amplitude modulation. A type of modulation where both the phase and the amplitude of the transmitted signal is varied in accordance with the information to be transmitted.

QPSK   Quadrature phase shift keying. A type of modulation where the phase is varied in accordance with the information to be transmitted and can take one of four discrete values.

Quantization error   The difference between the original waveform and the waveform after digitization has occurred.

RA   Radiocommunications Agency. The body in the United Kingdom responsible for the management of radio spectrum.

Rain Fading   The reduction in signal strength during heavy rain as a result of absorption of the radio signal by the moisture in the atmosphere.

RC   Raised cosine. A form of pulse shaping whereby a cosine segment is fitted to the leading and falling edges of the pulse.

RF   Radio frequency. The term for operations and components that act at the carrier frequency.

RFQ   Request for quotation. A document sent out, often by an operator, requesting quotations from suppliers.

RMS   Root mean square. A mathematical measure derived by taking the square root of the average of the squared input data.

RPE   Regular pulse excited. A form of AbS speech coding whereby the coder is excited by a regular stream of pulses.

RPE-LTP   Regular pulse excited—long-term prediction. A form of RPE speech coder where a long-term predictor is used to further remove the redundancy in the voice signal.

RS   Reed Solomon. The name of a type of block coding scheme.

RX-QUAL   Received signal quality. A measure of the quality of the received signal in a GSM system.

SACCH   Slow associated control channel. A channel within the GSM system used to send signaling information to and from the mobile while they are on a call.

Sampling rate   The frequency at which the input waveform is sampled.

SCH   Synchronization channel. A channel within the GSM control channels that is used to allow the mobiles to synchronize to the framing structure.

SCCP   Signaling connection and control part. A protocol within the SS7 signaling system used to provide a virtual connection and context for transmissions.

SCP   Service control point. An entity within an IN that controls the service provided to the users.

SDH   Synchronous digital hierarchy. A type of fixed network where all the switches are working in synchronization.

SELCALL   Selective calling. A type of signaling message sent at the start of a PMR broadcast that only alerts those for whom the call is destined.

SFN   Single frequency network. A technique used within digital TV where all the cells broadcast on the same frequency and a form of equalizer within the receiver removes any echoes.

SIB   Service-independent building block. A concept within IN whereby the components of the service offered to the users are identified at a level not specific to a particular network.

SIM   Subscriber Identity Module. A card within the GSM system that carries the identity of the subscriber. By being placed in a different phone it can allow subscribers to easily change phone but keep the same number.

SIR   Signal-to-interference ratio. The difference between a wanted signal and an interfering signal.

SME   Small- or medium-sized enterprise. This is typically defined as companies with less than 200 employees.

SMG   Special mobile group. The group within ETSI that conducts GSM standardization.

SMR   Specialized mobile radio. The name given to PMR within the United States.

SMS   Short message service. The ability within GSM to send text messages of up to 160 characters in length to a mobile.

SNR   Signal-to-noise ratio. The difference between the wanted signal and the noise.

SONET   Synchronous optical network. A network linked using optical fiber with all entities operating synchronously.

Squelch   A function within PMR of not allowing the signal to activate the loudspeaker if it falls below a certain level that is normally set by the user using a rotary knob.

SRBR   Short-range business radio. The name given by Motorola to a simple radio system that provides back-to-back communications.

SSP   Service switching point. An entity within an IN that provides the switching function.

SS7   Signaling system number 7. A protocol for sending information over a fixed network.

Switching in space   The addition of switching capability to a satellite system so that calls can be passed from one satellite to another until they are over the Earth station closest to their destination.

SWOT   Strengths, weaknesses, opportunities, and threats. An acronym used by managers when brainstorming strategy in order to consider all important issues.

TACS   Total Access Communications System. An analogue cellular radio system widely deployed within Europe.

TC   Technical committee. A committee set up within ETSI to study a particular issue.

TCAP   Transaction capability application part. Part of the SS7 protocol that links separate messages together into a context so that discrete responses can appear to be answers to a specific request.

TCM   Trellis Code Modulation. An advanced form of encoding whereby modulation and error correction are linked together in a means that enhances the information rate that can be transmitted.

TCP/IP   Transport control protocol/Internet protocol. The protocol used for transmission over the Internet.

TD-CDMA   Time division—code division multiple access. A combination of TDMA and CDMA.

TDD   Time division duplex. The transmission of uplink and downlink on the same frequency but at different times.

TDMA   Time division multiple access. A multiple access technique whereby users are given access to all the bandwidth but for a limited time.

TETRA   Terrestrial trunked radio. A standard for digital PMR radio systems.

TIA   Telecommunications Industry Association. The U.S. body responsible for telecommunications standardization of most systems, with the exception of GSM1900.

TMN   Telecommunications management network. A concept covering all fixed and mobile telecommunications networks whereby the same OMC can be used across a wide range of equipment by standardizing the OMC interface.

TRAU   Transcoder and Rate Adaption Unit. The entity within GSM responsible for transcoding voice and for adapting the rate of data to be transmitted, if required.

Trombone   A name for the route followed by some international calls within GSM where the call passes to the home country and back again in a long loop.

Trunking gain   The gain in efficiency of use of radio channels that results when a large number of channels are placed together in a pool.

TRX   Transmitter receiver. The name within GSM for the entity within the base station responsible for transmitting and receiving the signal.

TUP   Telephone user part. The part of the SS7 protocol responsible for simple speech communications.

UHF   Ultra High Frequency. The use of radio frequencies between 300 MHz and 3 GHz.

UIC   Union Internationale des Chemins de Fer (European Railway Union). The official body responsible for the harmonization of railway systems across Europe.

UMTS   Universal mobile telephone service. The European concept for the next generation of mobile radio systems to be deployed around the year 2002 onward.

USP   Unique selling point. A management term for the unique feature offered by their product or service.

VBS   Voice broadcast service. The addition to the GSM standard to provide broadcast capabilities.

VDSL   Very high-speed digital subscriber line. A form of DSL that will operate over distances of less than 100m but provide data rates of up to 50 Mbps in both directions.

VGCS   Voice group call service. The addition to the GSM standard to provide group calls.

VLR   Visitors' location register. The entity within GSM that is coupled to each MSC and stores the location of each mobile to the accuracy of a location area.

VSAT   Very small aperture terminal. A satellite terminal that communicates with geostationary satellites and has a dish diameter of less than 2m. Most VSATs are much smaller than this and can fit into briefcases.

W-CDMA   Wideband code division multiple access. A form of CDMA where the bandwidth of the transmitted signal is around 5 MHz.

Winner's curse   A problem that can occur in auctions whereby the winner has overvalued the market and subsequently goes bankrupt.

WLL Wireless local loop. A concept whereby fixed phones are provided using radio transmission.

xDSL Belonging to the family ADSL, HDSL, and VDSL.

XOR Exclusive OR mathematical operation.

About the author

William Webb graduated from Southampton University with a first class honors degree in electronic engineering and all top year prizes in 1989. In 1992, he earned his Ph.D. in mobile radio and in 1997 was awarded an MBA, all from Southampton University, United Kingdom.

From 1989 to 1993, William worked for Multiple Access Communications Ltd. as the Technical Director in the field of hardware design, modulation techniques, computer simulation, and propagation modeling. In 1993, he moved to Smith System Engineering Ltd., where he was involved in a wide range of tasks associated with mobile radio and spectrum management including devising the U.K. spectrum pricing proposals and writing the modifications to the GSM standard to introduce group and broadcast calls. In 1997, he moved to Netcom Consultants where he led the Wireless Local Loop division, and in 1998 he moved to Motorola where he is responsible for the Wireless Local Loop and In-Building products.

Dr. Webb has published over fifty papers, holds four patents, was awarded the IERE Premium in 1994, is a member of the IEE where he sits on the Editorial Advisory Panel, and is a senior member of the IEEE. He is the coauthor, with L. Hanzo, of *Modern Quadrature Amplitude Modulation* (Wiley and Sons, New York, 1992) and the author of *Introduction to Wireless Local Loop* and *Understanding Cellular Radio*, both published by Artech House, 1997.

Index

The Artech House Mobile Communications Series

John Walker, Series Editor

Understanding Digital PCS: The TDMA Standard,
Cameron Kelly Coursey

Understanding GPS: Principles and Applications, Elliott D. Kaplan,
editor

Vehicle Location and Navigation Systems, Yilin Zhao

Wideband CDMA for Third Generation Mobile Communications,
Tero Ojanperä, Ramjee Prasad

Wireless Communications for Intelligent Transportation Systems,
Scott D. Elliott, Daniel J. Dailey

*Wireless Communications in Developing Countries: Cellular and
Satellite Systems,* Rachael E. Schwartz

Wireless Data Networking, Nathan J. Muller

Wireless: The Revolution in Personal Telecommunications,
Ira Brodsky

For further information on these and other Artech House titles,
including previously considered out-of-print books now available
through our In-Print-Forever™ (IPF™) program, contact:

Artech House	Artech House
685 Canton Street	46 Gillingham Street
Norwood, MA 02062	London SW1V 1AH UK
781-769-9750	+44 (0) 171-973-8077
Fax: 781-769-6334	Fax: +44 (0) 171-630-0166
Telex: 951-659	Telex: 951-659
e-mail: artech@artech-house.com	e-mail: artech-uk@artech.house.com

Find us on the World Wide Web at: www.artech-house.com

gratia